An A to Z
of DNA
Science

What Scientists Mean When They
Talk about Genes and Genomes

An A to Z of DNA Science

What Scientists Mean When They Talk about Genes and Genomes

Jeffre L. Witherly

Galen P. Perry

Darryl L. Leja

Cold Spring Harbor Laboratory Press
Cold Spring Harbor, New York

An A to Z of DNA Science: What Scientists Mean When They Talk about Genes and Genomes

Developmental Editor	Catriona Simpson
Project Coordinator	Mary Cozza
Production Editor	Patricia Barker
Desktop Editor	Danny deBruin
Interior Designer	Denise Weiss
Cover designer	Ed Atkeson, Berg Design

Library of Congress Cataloging-in-Publication Data

Witherly, Jeffre L.
    An A to Z of DNA science : what scientists mean when they talk about genes
and genomes / Jeffre L. Witherly, Galen P. Perry, and Darryl L. Leja.
        p. cm.
    ISBN 0-87969-600-1 (paperback) (alk. paper); ISBN 0-87969-627-3 (hardcover)
    1. Genetics--Dictionaries.   2. DNA--Dictionaries.   3. Genes--Dictionaries.
4. Genomes--Dictionaries.   I. Perry, Galen P.
II. Leja, Darryl L. III. Title.
    QH427   .W58 2001
    576.5´03--dc21
                                                                        2001042138

10 9 8 7 6 5 4 3 2 1

All Cold Spring Harbor Laboratory Press publications may be ordered directly from Cold Spring Harbor Laboratory Press, 500 Sunnyside Blvd., Woodbury, NY 11797-2924. Phone: 1-800-843-4388 in Continental U.S. and Canada. All other locations: (516) 422-4100. FAX: (516) 422-4097. E-mail: cshpress@cshl.org. For a complete catalog of all Cold Spring Harbor Laboratory Press publications, visit our World Wide Web Site http://www.cshlpress.com.

Foreword

AS A SCIENTIST WHO SPEAKS OFTEN TO THE PUBLIC ABOUT THE POWER and promise of genetics, I frequently struggle to define certain terms that non-specialists find confusing. This problem has become even more acute now that the Human Genome Project has produced a map of our genetic blue-print, and the press has made "genes" such a popular topic. In addition, successes in genetic research are accelerating the pace of discovery in the broad field of biomedicine in general.

There is always difficulty in understanding technical terms, but the sit-uation is made worse in the field of genetic study because the field is exact-ing, the terminology specific and often uncommon, and the concepts often interrelated. The advantage that this book holds over many genetic glos-saries, indeed its core value, is the increased access into, and understanding of, the world of genetic research that it provides to the non-scientist through clear, well-defined terms. The approach, conversational in tone, peels away layers of scientific idiom while still maintaining clarity and accuracy. Helped by related term lists and clean, precise illustrations, the reader soon finds many quick and easy avenues toward a better under-standing of terms such as centimorgan, polymorphism, and cloning.

An A to Z of DNA Science is the best prescription I have found to help the general public better understand the common terms and concepts asso-ciated with modern genetics. I highly recommend it.

W. French Anderson

Dr. Anderson, a pioneer in the field of gene therapy, is Director of the Gene Therapy Laboratories at the University of Southern California School of Medicine.

Preface

COMMUNICATION IS MOST EFFECTIVE WHEN WORDS and images combine to convey a message in a clear, concise manner. This glossary is a product of such communication.

Genetic terminology, like any subject-specific language, can at times be confusing. This book is based on an audio version created in 1998 by the authors to take advantage of the new media tools and the limitless capacity to communicate via the Internet. That project, like this written edition, sought to provide a clearer understanding of complex genetic terms in a straightforward and informative manner.

Science communicators and genetic researchers have reviewed each entry for clarity and accuracy. Each term is the result of conversations with experts in genetics and related fields of study; cancer researchers were interviewed with regard to cancer terminology, genetic counselors helped explain ethical terms, and expert gene hunters assisted with the many terms surrounding the search for, and function of, genes.

This book is not intended to replace a course in genetics; its intent is to make easy the understanding of fundamental genetic terminology and principles. This is especially true in reference to helping the reader make sense of current news stories, conversations, medical opinions, and similar events in daily life that involve genetics.

We hope everyone who is challenged by genetic technology will find this book helpful.

Jeffre L. Witherly
Galen P. Perry
Darryl L. Leja

Acknowledgments

APPRECIATION GOES TO THE MANY GENETIC researchers at the National Human Genome Research Institute at the National Institutes of Health whose input, guidance, and support in developing learning tools about genetics helped make this book a reality. These include Melissa Ashlock, Michael Bittner, Leslie Biesecker, Bowles Biesecker, R. Michael Blaese, Lawrence Brody, Fabio Candotti, John Carpten, Settara Chandrasekharappa, Francis Collins, Eric Green, Ron King, Pu Paul Liu, Paul Meltzer, Lindsay Middelton, Richard Morgan, Maximilian Muenke, Robert Nussbaum, Donald Orlic, William Pavan, Mihael Polymeropoulos, Jennifer Puck, Thomas Ried, Danilo Tagle, Jeffrey Trent, Anthony Wynshaw-Boris, and Art Glatfelter.

Our thanks go also to Calvin Jackson, Lopa Basu, Stefanie Doebler, Judy Folkenberg, and Dan Hogan for help and advice in the formative stages of this project.

We also thank Catriona Simpson and Mary Cozza for their editorial advice and assistance, and Pat Barker, Denise Weiss, and Danny deBruin for the time and effort they applied to the design and production of the book.

Finally, a note of personal appreciation is owed to Jeffrey Trent, who shares our unwavering personal belief that genetic research must be made understandable to the average person.

Adenosine Deaminase (ADA) Deficiency

A severe immunodeficiency disease that results from a lack of the enzyme adenosine deaminase. It usually leads to death within the first few months of life.

> Adenosine deaminase, or ADA, is an enzyme inside certain cells that is an essential part of tissue metabolism. In particular, it is expressed in cells that provide for the development of the immune system. ADA deficiency, a very rare disease, occurs in children who are born with a defect in this gene. This defect leads to a disease known as severe combined immunodeficiency, or SCID. SCID usually causes the death of an affected child within the first few months of life from overwhelming infection. Historically, ADA is important because it was the first disease treated successfully with gene therapy.

Related terms: gene, gene therapy, severe combined immunodeficiency (SCID)

Adenoviruses

A group of DNA-containing viruses that cause respiratory disease, including one form of the common cold.

Adenoviruses have many properties of value to researchers. Scientists are learning how to use these viruses as transport mechanisms which, when manipulated in the laboratory, can be used for gene therapy or gene repair. Cystic fibrosis and cancer are potential target diseases for gene therapy.

Related terms: cancer, cystic fibrosis, DNA, gene therapy

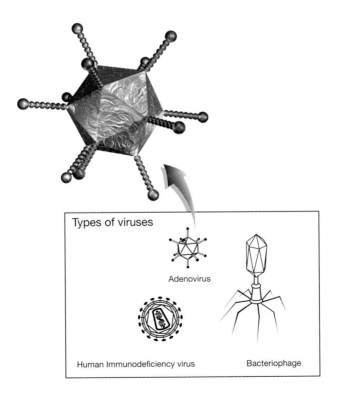

Types of viruses

Adenovirus

Human Immunodeficiency virus

Bacteriophage

Alagille Syndrome

A rare inherited liver disorder seen in infants and young children. The disease is characterized by a buildup of bile in the liver due to a deficiency or absence of normal bile ducts inside the liver and a narrowing of bile ducts outside the liver.

> Alagille syndrome comprises a wide range of abnormalities including problems of the liver, heart, eye, and vertebrae. The disease is initially identified by persistent jaundice in a newborn child, due to the absence of bile ducts in the liver. Individuals with Alagille syndrome are born with one copy of an important gene not functioning. Even though the remaining copy is normal, without both genes working properly, Alagille syndrome is the result.

Related terms: gene, inherited

Allele

One of the variant forms of a gene at a particular locus on a chromosome. Alleles express differently from person to person.

Different alleles produce variation in inherited characteristics such as hair or eye color and blood type. A single allele for each location is inherited separately from each parent. In an individual, one form of the allele, the dominant one, may be expressed more than another form, the recessive one. For example, whereas a given gene is responsible for eye color, the various forms, or alleles, determine which color.

Related terms: chromosome, deletion, DNA, dominant, gene, insertion, locus, recessive, substitution

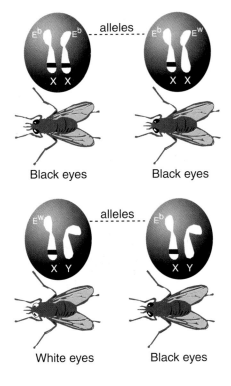

Amino Acids

A group of 20 different kinds of small molecules that link together in long chains to form proteins. Often referred to as the "building blocks" of proteins.

The 20 different amino acids are assembled in different combinations to make different proteins. For example, hair is made up of a specific protein, and that protein is made up of a specific sequence of amino acids. The muscle in your arm can be made up of the same 20 amino acids, but rearranged in a different sequence to produce muscle instead of hair. The body makes a series of very different proteins with different functions in this manner, by using the same 20 building blocks in different combinations.

Related terms: protein

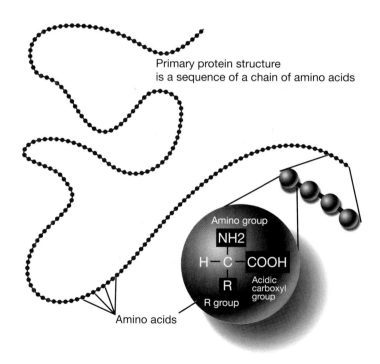

Primary protein structure
is a sequence of a chain of amino acids

Amino group
NH2
H — C — COOH
R
Acidic
carboxyl
group
R group

Amino acids

Animal Model

A laboratory animal useful in medical research because of specific characteristics that resemble a human disease or disorder. Scientists create animal models, usually laboratory mice, by transferring new genes into them.

The most common animal model is the mouse, which is thought to be nearly 90% genetically identical to humans. This similarity is extremely valuable to researchers because it enables them to identfy and study nearly all known diseases. Specific mouse models exist for numerous human diseases, including heart disease, diabetes, obesity, and cancer.

Related terms: deletion, gene, mouse model, mutation, transgenic

Antibody

A blood protein produced in response to an antigen, which then counteracts that antigen. Antibodies help the body develop an immunity to disease.

Antibodies are the foot soldiers of the human immune system. They are proteins that the body makes to help the immune system protect itself against invading foreign substances, such as bacteria.

Related terms: lymphocyte, protein

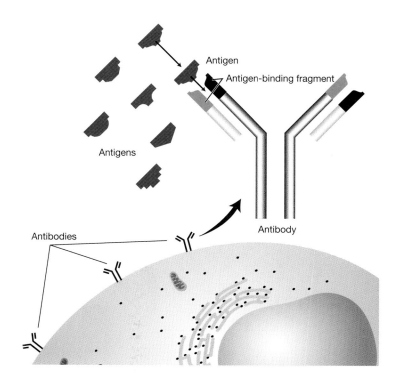

Antisense

The non-coding strand in double-stranded DNA. The antisense strand serves as the template for mRNA synthesis.

DNA, the chemical in a cell's nucleus that carries genetic instructions for making a living organism, has two strands and appears ladder-like. However, only one of the strands has information on it. The strand containing information is referred to as the coding or sense strand, whereas the other strand, its mirror image, is referred to as the antisense strand.

Related terms: cell, DNA, gene expression, messenger RNA (mRNA)

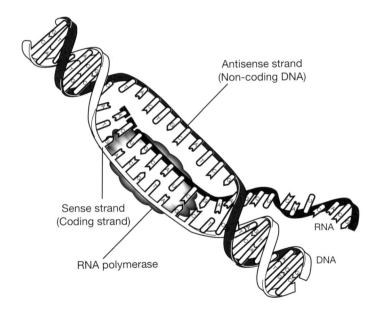

Antisense strand
(Non-coding DNA)

Sense strand
(Coding strand)

RNA polymerase

RNA

DNA

Apoptosis

Programmed cell death.

> *Apoptosis is the process by which damaged or unwanted cells are programmed to die when they are no longer needed by the body; this is often referred to as programmed cell death.*

Related terms: cell

Autosomal Dominant

A pattern of Mendelian inheritance whereby an affected individual possesses one copy of a mutant allele and one normal allele. Individuals with autosomal dominant diseases have a 50/50 chance of passing the mutant allele, and hence the disorder, on to their children

> *Autosomal dominant refers to a pattern of how we inherit traits or health conditions from our parents. Autosomal dominant conditions affect both genders; males and females are equally likely to have this trait or condition. For example, Huntington's disease is passed from parent to child in this manner.*

Related terms: autosome, dominant, gene, Huntington's disease, inherited, Mendelian inheritance, phenotype

Autosome

Any chromosome other than a sex chromosome. Humans have 22 pairs of autosomes.

Humans have 46 chromosomes. They come in 23 pairs. Of these 23 pairs, 22 are identical in women and men. The 23rd pair, known as the sex chromosomes (X and Y), account for the difference between men and women. The 22 chromosomes that are not different between men and women are called the autosomes.

Related terms: chromosome, sex chromosomes

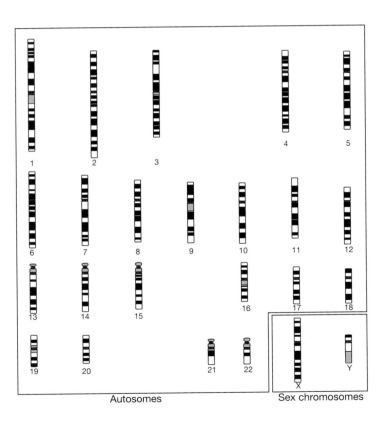

Autosomes   Sex chromosomes

Bacteria

Single-celled organisms. Bacteria are found throughout nature and can be beneficial or pathogenic to humans.

Bacteria are one of the simplest forms of life. Each bacterium contains in its cell all the information it needs to grow and replicate. Bacteria are found throughout nature and are both on and in our bodies all of the time.

Related terms: cell

Bacterial Artificial Chromosome (BAC)

Large segments of DNA (100,000 to 200,000 bases) from another species cloned into bacteria. Once the foreign DNA has been cloned into the host bacteria, many copies of it can be made.

A bacterial artificial chromosome— or a BAC as scientists call them for short—is a large piece of DNA genetically engineered to contain the components necessary for it to multiply as an artificial chromosome in bacteria. In this way, large individual fragments of human DNA can be cloned in bacteria. The cloned DNA in a BAC is typically smaller than that in its yeast counterpart, the yeast artificial chromosome, or YAC. BACs offer a number of important advantages to researchers, including ease of use in certain types of laboratory studies. BACs are used extensively for constructing high-resolution physical maps of human chromosomes and were particularly valuable in establishing the complete sequence of the human genome.

Related terms: base pair, chromosome, cloning, DNA, genome, yeast artificial chromosome (YAC)

Base Pair

Two bases that form a "rung of the DNA ladder." A DNA nucleotide is made of a molecule of sugar, a molecule of phosphoric acid, and a molecule called a base. The bases are the "letters" that spell out the genetic code. In DNA, the code letters are A, T, G, and C, which stand for the chemicals adenine, thymine, guanine, and cytosine, respectively.

DNA molecules are shaped like a twisting ladder. DNA consists of two strands that wrap around one another forming this twisted ladder. Each rung of the ladder connects the two strands with a pair of bases, or a base pair. Bases are also often referred to as nucleotides. These bases, A, C, G, and T, follow certain rules on how they can pair up on each rung. Each base on a strand of DNA pairs specifically with another base on the opposite strand of DNA to form the rung of the ladder. Adenine always pairs with thymine, and guanine always pairs with cytosine. Base pairs are frequently used as a measure of the length of a piece of DNA. For example, a piece of DNA might be 500 base pairs long. That means that it consists of two strands of DNA, each of which contains 500 nucleotides that are paired with the 500 nucleotides on the other strand to form 500 base pairs.

Related terms: DNA, genetic code, nucleotide

Deoxyribonucleic Acid (DNA)

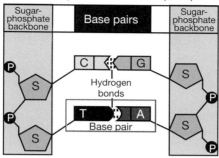

Cytosine   Guanine

Thymine   Adenine

Birth Defect

A structural, functional, or metabolic abnormality present at birth that results in physical or mental disability or fatality. There are more than 4,000 known birth defects, which may be caused by genetic or environmental factors. Each year in the United States, about 150,000 babies are born with birth defects.

The term birth defect generally refers to an abnormality in physical development that is present at birth, although most birth defects are present well before birth in the developing fetus. There are many different causes of birth defects and not all are genetic in origin. Some birth defects are caused by harmful aspects of the environment, such as exposure to certain chemicals, radiation, and similar damaging agents. Other birth defects are genetic and occur when a gene, or several genes, fail to function properly.

Related terms: enzyme, gene

Bone Marrow Transplantation

A medical procedure to replenish the blood-cell-producing soft tissue within bones. Transplants are necessary when marrow is destroyed by drug or radiation therapy for cancer.

Bone marrow transplantation is a clinical procedure used to restore normal bone marrow function in patients with blood diseases. Blood cell disorders occur as a result of exposure of bone marrow cells to toxic chemicals, high doses of radiation, and often because of genetic mutations. Examples of genetic disorders include sickle cell anemia and adenosine deaminase deficiency. One approach to correct blood cell diseases involves the transplantation of healthy, closely matched bone marrow into the patient. Often the donor bone marrow is from a sibling or parent. The recipient is generally prepared for the transplant procedure by treatment with chemicals and/or total body radiation in an effort to destroy the diseased bone marrow. If the procedure is successful, the hematopoietic stem cells in the transplant will repopulate the patient's bone marrow, enabling the body to once again produce normal blood cells. Recently, clinicians have been able to transplant highly purified hematopoietic stem cells harvested from circulating blood and umbilical cord blood. These sources of stem cells provide bone marrow transplants containing fewer, or no, cells that will react negatively with the recipient's tissue.

Related terms: cancer, hematopoietic stem cells, leukemia, mutation

BRCA1 / BRCA2

The first breast cancer genes to be identified. Mutated forms of these genes are believed to be responsible for about half the cases of inherited breast cancer, especially those that occur in younger women. Both are tumor suppressor genes.

Breast cancer ranks second only to lung cancer as a cause of cancer deaths in women. There are several genes known to lead to an inherited susceptibility to breast cancer. Two of these that researchers know the most about are the BRCA1 and BRCA2 genes. A person (male or female) who is born with an altered copy of a BRCA1 or BRCA2 gene has a greatly increased risk of developing breast cancer.

Related terms: cancer, gene, mutation, tumor suppressor gene

Cancer

Diseases in which abnormal cells divide and grow unchecked. Cancer can spread from its original site to other parts of the body and can be fatal if not treated adequately.

Cancer is responsible for more than 300,000 deaths per year in the U.S. alone. Cancer affects people at any age, is usually chronic, and is often fatal. It is not surprising that a vast global research effort is focused on the study of cancer. Unfortunately, many compounding problems exist. There are many kinds of cancer, and a remarkable difference in the incidence of cancer as determined by sex, age, race, geography, and other factors. Cancerous tumors may occur at almost any part of the body, and there are many kinds of tumors. There are multiple causes for tumors, and multiple diseases caused by tumors. Therefore, it is difficult to form a unifying definition of this collection of diseases. However, if there is a unifying basis for cancer, it is that all cancers are composed of cells in which cell division has somehow gone awry. These cells form malignant tumors that may invade surrounding tissue, metastasize to other parts of the body, recur after attempted removal, and cause the death of the patient if not eliminated.

Related terms: carcinoma/sarcoma, cell

Candidate Gene

A gene, located in a chromosome region suspected of being involved in a disease, whose protein product suggests that it could be the cause of the disease in question.

> When researchers search for a gene associated with a genetic disease, they usually focus on candidate genes. A gene is considered to be a candidate gene either when its encoded protein represents a logical possibility for being involved in the disease, or when the gene is physically located in a region of the genome known to contain the diseased gene. The second situation frequently occurs in positional cloning research where a diseased gene is identified on the basis of its position within the genome. In a typical positional cloning project, a small area of the genome is first identified as containing the diseased gene. Next, all the genes residing in that region immediately become possible candidate genes. If something is known, or can be inferred, about the encoded protein of a given candidate gene, and if a convincing case can be made for how that protein may be involved in the disease, that gene would likely be considered a strong candidate gene. Research efforts would then focus on finding mutations in that candidate gene.

Related terms: chromosome, cloning, gene, positional cloning/functional cloning, protein

Carcinoma / Sarcoma

Any of the various types of cancerous tumors that form in the epithelial tissue, which is the tissue that forms the outer layer of the body surface and lines the digestive tract and other hollow structures. Examples of this kind of cancer include breast, lung, and prostate cancer.

> *Carcinoma and sarcoma are two different types of cancer named for the tissue where the malignancy occurred in the body. Carcinoma is cancer that has derived from either the ectodermal or endodermal epithelial cells. For example, prostate cancer is usually prostate carcinoma, because it is derived from the endodermal epithelial cells of the prostate gland. Other carcinomas include skin and breast cancer. Sarcoma is cancer that has derived from mesodermal epithelial cells. Leukemia is an example of this type of cancer and is known as leukocytic sarcoma.*

Related terms: cancer, prostate cancer

cDNA Library

A collection of DNA sequences generated from mRNA sequences. This type of library contains only protein-coding DNA (genes) and does not include any non-coding DNA.

The DNA in an organism's genome can be conceptually divided into two parts: that which encodes for proteins, called genes, and that which does not, called non-coding DNA. In the process of making a protein, the DNA in a gene is read and converted into RNA, which is another type of nucleic acid. This RNA serves as a messenger between the DNA and the protein-making machinery of the cell, and thus it is called messenger RNA or mRNA. The mRNA instructs the protein-making machinery to assemble amino acids in a precise order to make the encoded protein. It is also possible for researchers to convert the mRNA in a cell or tissue back into "complementary" DNA, or cDNA, and then clone it for further use. To do this, researchers use a specialized recombinant DNA library, called the cDNA library. cDNA libraries consist of clones containing the protein-coding portions of DNA — or in other words, the genes within a genome.

Related terms: amino acids, DNA, gene, genome, mRNA, protein, ribonucleic acid (RNA)

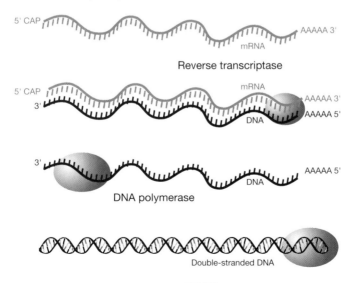

Cell

The basic unit of any living organism. It is a small, watery, compartment filled with chemicals and a complete copy of the organism's genome.

A cell is the fundamental unit of life. It is essentially the smallest self-contained unit of a living organism. Our body consists of billions of individual cells. Each of our cells is an individual entity bounded by a membrane and containing a single copy of the human genome stored in the nucleus. Not all cells have a nucleus; for example, your red blood cells don't have a nucleus, but most of the rest of the cells in the body do. This gives each cell the capability to, in principle, divide by duplicating its DNA. Thus, the cell is the fundamental unit of replication of the organism and carries with it all of the products of the genome needed to take care of the life of that cell.

Related terms: genome, nucleus

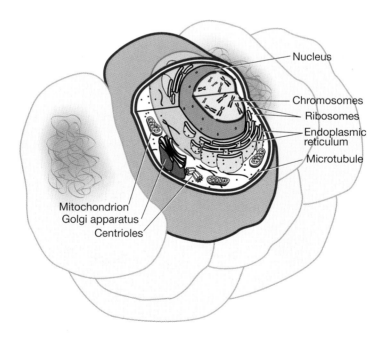

Nucleus

Chromosomes

Ribosomes

Endoplasmic reticulum

Microtubule

Mitochondrion

Golgi apparatus

Centrioles

Centimorgan

A measure of genetic distance that tells how far apart two genes are. Generally one centimorgan equals about 1 million base pairs.

A centimorgan is the standard unit by which geneticists describe how far apart two places, or two genes, are on a chromosome. A centimorgan, by definition, is the distance apart two loci are if there is a 1% chance that there will be a crossover between those two loci during meiosis—that is, in passing of that chromosome from parent to child. For humans, a centimorgan translates into about a million base pairs.

Related terms: chromosome, gene, locus

Centromere

The constricted region near the center of a human chromosome. This is the region of the chromosome where the two sister chromatids are joined to one another.

The centromere divides the chromosome into two arms or parts. The centromere's position on any given chromosome remains constant for that chromosome. The centromere is also the site of attachment of spindle fibers during mitosis and meiosis. It is the region where the two chromatid strands are held together on a chromosome.

Related terms: chromosome

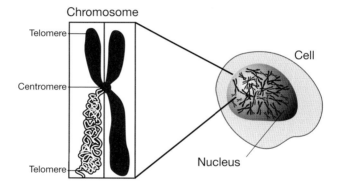

Chromosome

A thread-like packet of genes, and other DNA, in the cell's nucleus. Different kinds of organisms have different numbers of chromosomes. Humans have 23 pairs of chromosomes, 46 in all: 44 autosomes and 2 sex chromosomes (X and Y).

All living cells house their genetic blueprints in structures called chromosomes. Chromosomes consist of tightly packed DNA and proteins, and appear almost thread-like or rope-like when examined microscopically. In the case of cells containing a nucleus, like most human cells, the chromosomes are all located together within the nucleus. Different organisms have different numbers of chromosomes. Each parent contributes one chromosome to each pair, so children get half of their chromosomes from their mothers and half from their fathers. In this way a new genetic blueprint is created in each child, representing the mixture from each parent.

Related terms: autosome, birth defect, cell, DNA, gene, nucleus, protein, sex chromosomes

Chromosome *(Continued)*

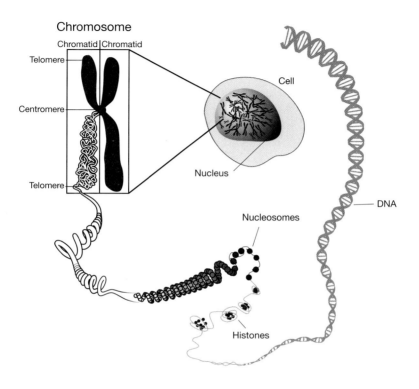

Chromosome

Chromatid | Chromatid

Telomere

Centromere

Telomere

Cell

Nucleus

Nucleosomes

Histones

DNA

Cloning

The process of making copies of a specific piece of DNA, usually a gene. When geneticists speak of cloning, they do not mean the process of making genetically identical copies of an entire organism.

A clone, and cloning, are terms that can have completely different meanings. When geneticists speak of clones or cloning, they are almost always talking about the process of making copies of a specific piece of DNA (usually a gene), and not the process of making genetically identical copies of an entire organism. Another way of defining a clone is that it is multiple copies of one piece of DNA. Researchers use the process of cloning a gene, or making copies of a gene, because it is a simple step to isolate one gene away from other genes, and then (using polymerase chain reaction) make many identical copies of that gene for further research. This allows researchers to study the effect a specific gene has in isolation, while ignoring the effect of other genes.

Related terms: DNA, gene, polymerase chain reaction

Codon

Three bases in a DNA, or RNA, sequence that specify a single amino acid.

Codon is a term that defines the words that DNA uses to specify the genetic code. DNA is composed of four bases: G, A, T, and C. These bases, or letters, encode all of the genetic information in an organism. The words that are used in this process are three-letter words, which are always composed of three of those four bases. These words can be translated into specific amino acids, which are the building blocks of proteins. Thus, a codon is a three-letter word specifying either the start of a protein, the end of a protein, or one of the amino acid building blocks.

Related terms: amino acids, DNA, genetic code, ribonucleic acid (RNA)

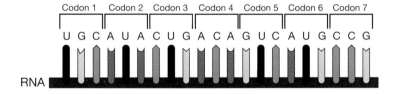

Congenital

Any trait or condition that exists from birth.

The term congenital refers to a physical or mental condition that is evident at, or even before, birth. Congenital conditions may be inherited as part of the parent's genetic contribution to the child, or they may be the result of an insult to the fetus during pregnancy. The term congenital is often associated with the words defect or abnormality — as in congenital defect or congenital abnormality. However, congenital simply describes a condition present at birth, regardless of how it was caused.

Related terms: birth defect

Contig

A chromosome map showing the locations of those regions of a chromosome where contiguous DNA segments overlap.

The physical mapping of DNA often involves isolating large pieces of DNA in clones, such as BACs and YACs. These clones are then analyzed to determine which ones contain DNA in common, or in other words, overlap. A collection of overlapping clones that together contains a "contiguous" segment of the starting DNA is called a contig. Contigs provide the ability to study a complete and often large segment of the genome using a series of overlapping clones. This is particularly important when searching for genes of interest and for sequencing large structures of a chromosome. A contig map is a physical map reflected by an overlapping series of clones.

Related terms: bacterial artificial chromosme (BAC), chromosome, cloning, DNA, gene, genome, physical map, yeast artificial chromosome (YAC)

Cystic Fibrosis

A hereditary disease whose symptoms usually appear shortly after birth. They include faulty digestion, breathing difficulties and respiratory infections due to mucus accumulation, and excessive loss of salt in sweat.

Cystic fibrosis is the most common, potentially fatal, recessive disease in Caucasians, affecting nearly one in 25 newborns. Cystic fibrosis is a chronic and progressive genetic disease of the mucous glands. It primarily affects the respiratory and digestive systems in children and young adults. The prognosis of cystic fibrosis has improved over the last 40 years, but still the average survival is only to about age 30. In 1989, the gene for cystic fibrosis was found on chromosome 7 using positional cloning techniques. This discovery resulted in new ideas about ways to treat the disease — for example, with gene therapy and with drug therapy. Many researchers share optimism and enthusiasm that this will be one of the first successes of the new genetic medicine in terms of curing a potentially fatal disease, although this breakthrough may not occur in the near future.

Related terms: gene, gene therapy, positional cloning, recessive

Cytogenetic Map

The visual appearance of a chromosome when stained and examined under a microscope. Particularly important are visually distinct regions, called light and dark bands, which give each of the chromosomes a unique appearance that allows a chromosome to be studied by karyotyping.

A cytogenetic map shows the visual appearance of a chromosome when properly stained in the laboratory and examined microscopically. Particularly important parts of a cytogenetic map are the distinct regions called bands that stain either lightly or darkly, and thus are called light bands or dark bands. Each band contains approximately 5 to 10 million bases of DNA. In the case of the human genome, this banding pattern gives each of the chromosomes a unique appearance. This feature allows a person's chromosomes to be studied with a clinical test known as a karyotype. Karyotyping allows a visual surveying of the genetic map of an individual for visually detectable alterations such as chromosome gains or losses, as well as deletions, rearrangements, and other major structural changes.

Related terms: chromosome, deletion, karotype

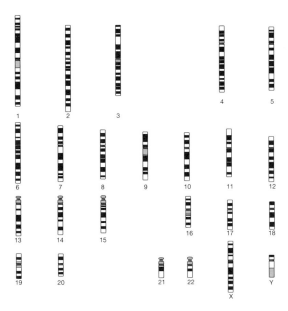

Deletion

A particular kind of mutation: loss of a piece of DNA from a chromosome. Deletion of a gene or part of a gene can lead to a disease or abnormality.

There are many different kinds of mutations that may cause genetic changes or diseases. One of them is a deletion. A deletion is actually just what it sounds like— a portion of the genome is missing, or deleted. This may be caused by an event such as a chemical mutagen or irradiation, or just by chance during cell division. Whatever the cause, a piece of DNA becomes deleted from a chromosome. If this deleted portion includes a gene or some other piece of DNA required for normal cell function, the consequence can be negative for the organism and result in disease.

Related terms: chromosome, DNA, gene, genome, mutation

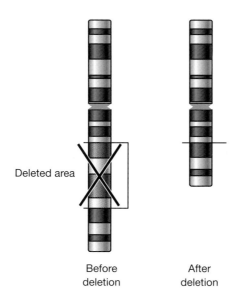

Deleted area

Before
deletion

After
deletion

Diabetes Mellitus

A highly variable disorder in which abnormalities in the ability to make and/or use the hormone insulin interfere with the process of turning dietary carbohydrates into glucose, the body's fuel. There are two types of this disease: Type I is known as insulin-dependent diabetes mellitus, and type II is known as non-insulin-dependent diabetes mellitus.

Type I diabetes (also called insulin-dependent diabetes mellitus or juvenile onset diabetes mellitus) affects primarily children and young adults. This type of diabetes is characterized by an autoimmune process that destroys the insulin-producing cells in the pancreas. Type II diabetes (also called NIDDM for non-insulin-dependent diabetes mellitus or adult onset diabetes mellitus) is also a problem of glucose and insulin metabolism, but it is a very different disease. People with type II diabetes tend to have insulin resistance. Their body is making insulin, but it isn't working the way it should work. Eventually, they also will have insulin deficiency. NIDDM has strong genetic underpinnings, but it doesn't follow the principles of inheritance that Gregor Mendel would have approved of—it's not dominant, it's not recessive, it's not sex-linked; instead, it appears to be polygenic. There are multiple genes involved; a combination of the inheritance pattern of multiple genes, plus environmental factors (particularly obesity), determines whether or not a person will develop the disease.

Related terms: dominant, gene, inherited, mapping, recessive

Diploid

The number of chromosomes in most cells, except the gametes. In humans, the diploid number is 46.

> *Diploid describes the complete number of copies of the genome in a given cell. For example, normal human cells contain two copies of the genome, one obtained from each parent; those cells are referred to as diploid cells (di- meaning two, and ploidy referring to the number of copies). In contrast, haploid is the number of copies of the genome in the mature reproductive cells of the body — the eggs and the sperm. Haploid means half the number of copies in the diploid genome.*

Related terms: cell, chromosome, genome, haploid

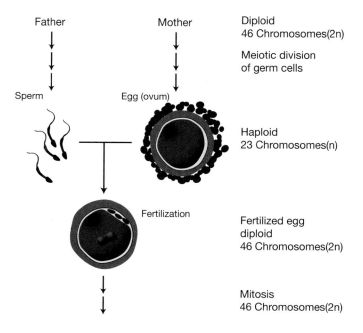

Father

Mother

Diploid
46 Chromosomes(2n)

Meiotic division
of germ cells

Sperm

Egg (ovum)

Haploid
23 Chromosomes(n)

Fertilization

Fertilized egg
diploid
46 Chromosomes(2n)

Mitosis
46 Chromosomes(2n)

DNA

The chemical inside the nucleus of a cell that carries the genetic instructions for making living organisms.

DNA, or deoxyribonucleic acid, is the chemical inside the nucleus of our body's cells that carries the genetic instructions for making a human. DNA was only recently observed to be the genetic material of the cell. Before the 1950s, it was commonly thought that proteins carried the genetic information of all cells. DNA is also referred to as the double helix.

Related terms: cell, double helix, nucleus, protein

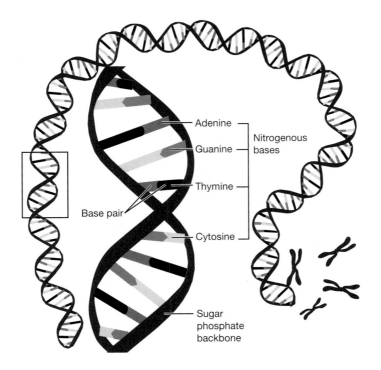

DNA Replication

The process by which the DNA double helix unwinds and makes an exact copy of itself.

Amazingly, if you were able to open the nucleus of one of your body's cells and fold out the DNA inside, you'd find each cell contains almost three feet of DNA. This three feet of DNA is made up of over three billion different components. Every time the cell divides, it has to make a copy of all three feet of DNA. The cell accomplishes this in about half an hour. This process is called DNA replication. Once the DNA is replicated, the cell will split into two cells, each of which has a copy of the original DNA. Occasionally, during replication, mistakes are made. The cell has a very sophisticated process of finding mistakes and repairing them. If the mistakes are not repaired, they can potentially cause the cell to die, lead to tumor formation, or cause other harmful results.

Related terms: cell, double helix

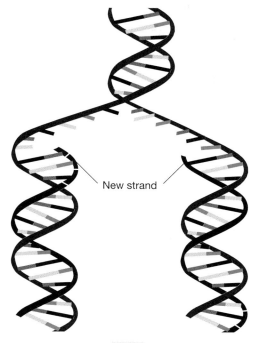

New strand

DNA Sequencing

Determining the exact order of the base pairs in a segment of DNA.

DNA consists of a linear string of four bases or letters: G, A, T, and C. It is the order of these bases that encodes the important information in the genetic blueprint of an organism. DNA sequencing is the process whereby researchers actually determine the precise order of bases along a stretch of DNA. This typically involves subjecting DNA fragments to gel electrophoresis, which results in the separation of DNA fragments that differ in size by a single base. Each fragment is then detected due to the presence of either an attached radioactive or a fluorescent tag. The most commonly used instruments for DNA sequencing use a laser beam to detect fluorescent-tagged DNA. From the pattern of the resulting DNA fragments, the underlying DNA sequence can be deduced.

Related terms: electrophoresis, genetic code

Dominant

A gene that almost always leads to a specific physical characteristic; a dominant gene predicates a 50/50 chance in each pregnancy for passing on a disease to one's children.

In the study of the inheritance of traits in families it was observed, as far back as the time of Gregor Mendel, that certain traits could be inherited from generation to generation if one parent of the offspring had the trait. This type of trait is called a dominant trait. Dominant means all that is required for that trait to be inherited from one generation to the next is that the parent who has the trait passes on the gene, or genes, for that trait to the child. This result is independent of the genes that are passed on by the other parent.

Related terms: gene, genome, recessive

Double Helix

The structural arrangement of DNA.

The double helix structure of DNA was determined in 1953 by James Watson and Francis Crick who, while studying X-ray photographs, were able to deduce the actual chemical structure of the inherited molecule that carries all the genetic information necessary to make a human being. The double helix is ladder-like in shape, and each ladder rung consists of a pair of bases. It is the sequence of these base pairs that determines the instructions for a particular organism. Human DNA contains about 3 billion base pairs. All mammalian DNA is approximately the same length, but in simpler organisms the DNA is usually shorter. For instance, the fruit fly only has about 150 million base pairs, the roundworm about 100 million, yeast about 15 million, and bacteria between 1 and 5 million, depending on the organism.

Related terms: base pair, DNA, nucleotide

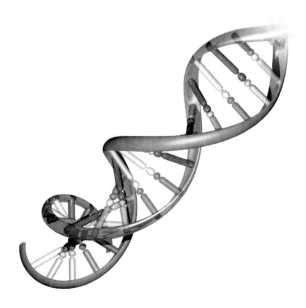

Duplication

A particular kind of mutation: production of one or more copies of any piece of DNA, including a gene or even an entire chromosome

> *Many different kinds of DNA alterations, or defects, cause genetic diseases; one is called duplication. Duplication occurs when a portion of the genome that is normally present in two copies is present in more than two copies. When this happens, the multiple copies of DNA create an increased amount of gene product from that gene. The extra amount of gene product, or protein, triggered by the mutation causes a specific disease.*

Related terms: cancer, chromosome, DNA, gene, genome, mutation, phenotype, protein

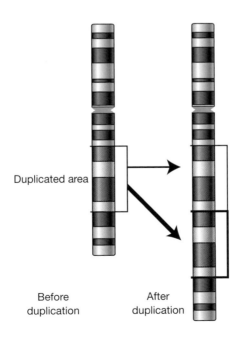

Duplicated area

Before duplication

After duplication

Electrophoresis

A process in which molecules can be separated according to size and electrical charge. Application of low-level current forces the molecules through pores in a thin layer of gel, with smaller fragments traveling farther than large ones. The process is sometimes called gel electrophoresis.

> Electrophoresis is a laboratory process that separates DNA and protein molecules. The separation occurs on what is known as a gel, which is similar to household gelatin. The gel is poured in a sheet and the DNA or protein mixture is added at one end. Electricity is then applied across the gel, forcing the molecules to separate. The smaller molecules will sink more quickly because they move through the gel more easily. The larger DNA molecules stay nearer the top, which makes viewing those specific molecules easier.

Related terms: DNA, protein, ribonucleic acid (RNA)

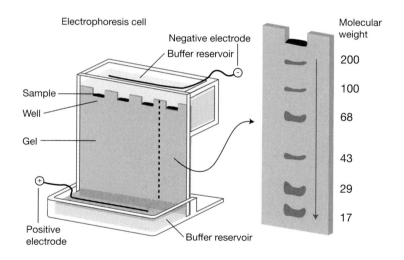

Enzyme

A protein that encourages a biochemical reaction, usually speeding it up. Organisms could not function without enzymes.

Enzymes are proteins that carry out essentially every chemical reaction in the human body. Many of the reactions that occur during normal metabolism and normal function of the human body would not occur without enzyme catalysis. Enzymes usually accelerate the speed with which these reactions would occur.

Related terms: protein

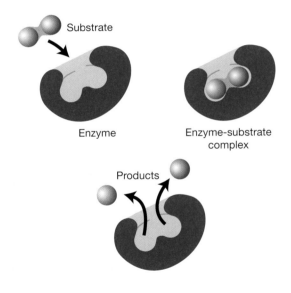

Substrate

Enzyme

Enzyme-substrate complex

Products

Exon

The region of a gene that contains the code for producing the gene's protein. Each exon codes for a specific portion of the complete protein. In some species (including humans), a gene's exons are separated by long regions of DNA (called introns or sometimes "junk DNA") that have no apparent function.

> *When defining exon, it is also best to define intron, since both are closely associated in organisms with one or more cells. In short, genes are divided into two sections, the exon and the intron. The exon represents that part of the gene that actually codes for a protein during translation, whereas introns act as spacers or stretches of DNA between each exon within a gene. The introns are then removed during events that occur within RNA molecules, allowing the exons to join together to form mRNA (messenger RNA).*

Related terms: gene, genetic code, messenger RNA, protein

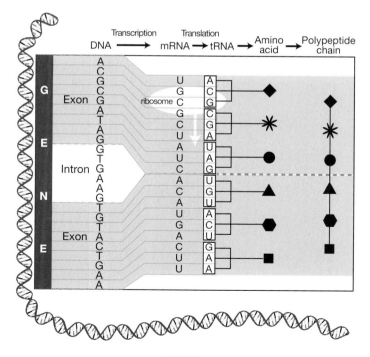

Fibroblast

A type of cell found just underneath the surface of the skin. Fibroblasts are part of the support structure for tissues and organs.

> Fibroblasts are flat, square-shaped connective tissue cells. Fibroblasts form the scaffolding, or structure, around tissues and organs in the bodies of all vertebrates.

Related terms: cell

Fluorescence In Situ Hybridization (FISH)

A process that vividly paints chromosomes or portions of chromosomes with fluorescent molecules. This technique is useful for identifying chromosomal abnormalities and gene mapping.

> *FISH is the acronym for fluorescence in situ hybridization. This laboratory technique allows the visualization of genes and chromosomes during all stages of the cell cycle. The process involves the fluorescent labeling of DNA probes, followed by a hybridization step that results in a multicolored array of chromosomes allowing visualization of the sequence using fluorescent microscopy. FISH is used for gene mapping studies and for the analysis of chromosomal aberrations, including those that occur in cancer cells.*

Related terms: chromosome, gene, hybridization, mapping

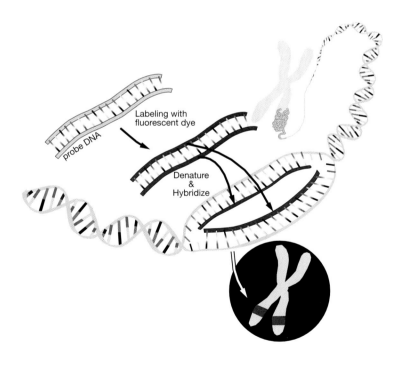

Fragile X Syndrome

The disorder is one of a group of diseases that results from an unusual kind of mutation: a repeating sequence of three letters of the DNA code, called a triplet repeat or trinucleotide repeat. In Fragile X, the repeating triplet is CGG, cytosine-guanine-guanine, in a gene on the X chromosome. The larger the number of repeats an individual possesses, the more likely he or she is to be seriously impaired. Individuals with a few repeats are carriers but often are not affected by the disease.

> *Fragile X syndrome is a genetic disorder that results in mental impairment or mental retardation. It is the second most frequent developmental cause of mental retardation in humans and can cause learning problems in both males and females. Fragile X syndrome is the result of a gene defect that causes an expansion of three nucleotides in the part of the gene that regulates gene expression. When this group of nucleotides expands to more than 200 copies, expression of the gene stops (or turns off), and the result is Fragile X syndrome. The syndrome name stems from the very fragile and thin appearance of the X chromosome when studied under a microscope.*

Related terms: cell, chromosome, gene, nucleotide, sex chromosomes

Gene

The functional and physical unit of heredity passed from parent to offspring. Genes are pieces of DNA, and most genes contain the information for making a specific protein.

A gene is a stretch of DNA that contains the instructions for making, or "coding for," a particular protein. The DNA that constructs a gene resides in the nucleus of the cell and serves as the storehouse of information, but it doesn't actually perform the work that accomplishes its given instruction; the proteins do that. Genes may be as short as 100 base pairs or as long as a couple of million base pairs.

Related terms: base pair, DNA, genome, nucleus, protein

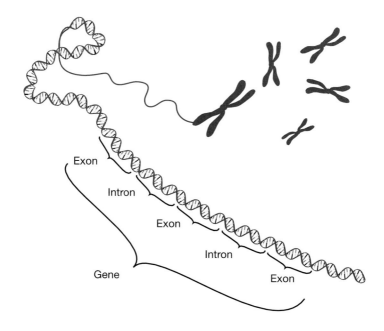

Gene Amplification

An increase in the copy number for any particular piece of DNA. A tumor cell amplifies, or copies, DNA segments naturally as a result of cell signals and, sometimes, environmental events.

Normal cells, when they're ready to divide, copy their DNA. Tumors develop in the body when a cell has problems, or makes a mistake, when copying its DNA. Occasionally, a cell might make a mistake and copy portions of its DNA numerous times; this is called gene amplification. If the region that is copied happens to contain an oncogene, this may lead to that cell becoming a tumor. In the case of certain tumors, specific genes are copied thousands of times. The number of extra, or unwanted, copies determines the aggressiveness, or ability of the tumor to spread.

Related terms: cell, DNA, gene, oncogene

Gene Expression

The process by which proteins are made from the instructions encoded in DNA.

Gene expression describes the protein production process carried out by instructions contained in DNA. Each cell contains a blueprint to build the whole body. Proteins are the basic building blocks of the body. In some respects, the human blueprint is similar to a blueprint for a whole city. If you think in those terms, you wouldn't construct a factory in a family neighborhood, and you wouldn't locate a busy airport next to the elementary school. Similarly, in the cell, if a cell is in your skin, you don't want that cell to express hemoglobin, which transports oxygen in the blood, even though the complete blueprint in each cell has the instructions to do this. Gene expression provides control over the body's blueprint by selectively producing specific types of proteins in our cells.

Related terms: DNA, gene, protein, messenger RNA (mRNA)

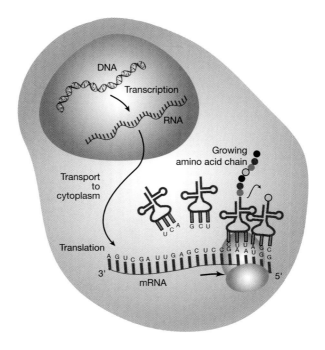

Gene Mapping

Determining the relative positions of genes on a chromosome and the distance between them.

> *Gene mapping can be done using genetic or linkage mapping, which determines a statistical distance between two genes. Alternatively, it can be done by physical mapping, which determines the distance, by nucleotides, or base pairs, of DNA, between two genes. While both mapping methods are useful, they are usually done in the order of linkage mapping first, and physical mapping second, particularly in the process of positional cloning or gene isolation of inherited human diseases.*

Related terms: base pair, chromosome, DNA, gene, genome, linkage, nucleotide, physical map

Gene Pool

The sum total of genes, with all their variations, possessed by a particular species at a particular time.

> *The human body contains somewhere between 30,000 and 35,000 genes. Each of these genes has a specific location on a particular one of the body's 23 chromosomes. For any particular gene, there may be variants, or alleles. The gene pool is the sum total of all of the variants of a particular gene across the entire population.*

Related terms: allele, chromosome, gene

Gene Therapy

An evolving technique used to treat inherited diseases. The medical procedure involves either replacing, manipulating, or supplementing nonfunctional genes with healthy genes.

> *The simple definition of gene therapy is the use of information stored in our genes for the treatment of human disease. When scientists realized there was so much information stored in our gene pool and genome, they began to think of ways to use it for the treatment of inherited or genetic diseases. Early attempts to use gene therapy involved the concept of correcting a defective gene by putting in DNA, genomic DNA, or the cDNA that codes for the specific defective protein and adding it to the cells of a patient to restore that cell to normal function. Examples of this idea include placing clot-preventing genes into vascular grafts used in coronary bypass surgery, or genetically modifying a T cell so that an HIV virus can no longer enter that cell or divide.*

Related terms: cDNA library, gene, genome, inherited, protein

Genetic Code

The instructions housed in genes that tell the cell how to make a specific protein. A, T, G, and C are the "letters" of the DNA code; they stand for the chemicals adenine, thymine, guanine, and cytosine, respectively, that make up the nucleotide bases of DNA. Each gene's code combines the four chemicals in various ways to spell out 3-letter "words" that specify which amino acid is needed at every step in making a protein.

DNA is a language that tells a cell how to make all of the proteins that cell must produce to function normally. This language is composed of A, C, G, and T. These letters are the four bases that make up the strands of DNA. The genetic code allows the cell to produce this string of four base letters into words and interprets these words as to how to make particular proteins. The words are all the letters in a given length of DNA, and they specify which amino acid is needed during the making of a protein. For example, the word TTT spells the amino acid phenylalanine, whereas TTA spells leucine, and GTA spells valine. Each of the twenty amino acids has its own particular word set that constitutes the genetic code for all of those amino acids.

Related terms: amino acids, base pair, DNA, gene, nucleotide, protein

Genetic Counseling

A short-term educational counseling process for individuals and families who have a genetic disease or who are at risk for such a disease. Genetic counseling provides patients with information about their condition and helps them make informed decisions.

> *Genetic counseling provides information to support afflicted individuals and families struggling to understand a genetic condition, or a risk for a condition. Trained professionals with backgrounds in health-care delivery, genetics, ethics, and research generally lead the counseling sessions. Areas of discussion include reproductive choices, including risks and options, as well as the potential consequences of testing.*

Genetic Map

A chromosome map of a species that shows the position of its known genes and/or markers relative to each other, rather than as specific physical points on each chromosome.

> *A genetic map, also known as a linkage map, represents the positions of genetic markers along a stretch of DNA. Genetic markers reflect DNA sequences that differ among individuals. A genetic marker, also known as a polymorphism, can range from sequence differences that result in identifiable phenotypes, or characteristics, to more innocent sequence differences that have no noticeable effect on an individual. Such sequence differences are scattered throughout our DNA and serve as the basis for building detailed genetic maps of any genome, including human. Genetic maps are essential to numerous applications, one example being the search for genes associated with a disease by a positional cloning.*

Related terms: chromosome, gene, genetic marker, polymorphism, positional cloning/functional cloning

Genetic Marker

A segment of DNA with an identifiable physical location on a chromosome, and whose inheritance can be followed. A marker can be a gene, or it can be some section of DNA with no known function. Because DNA segments that lie near each other on a chromosome tend to be inherited together, markers are often used as indirect ways of tracking the inheritance pattern of a gene that has not yet been identified, but whose approximate location is known.

A genetic marker is a segment of DNA that has a known particular location on a chromosome, yet differs between individual families. This subtle change in DNA sequence acts as a flag passed on from generation to generation, and can be detected only through laboratory tests. For example, children separated at birth could be tested, and on the basis of certain markers, it could be determined whether they were from the same family, or for that matter, whether their offspring were from the same family.

Related terms: chromosome, DNA, gene, inherited

Genetic Screening

Testing a population to identify a subset of individuals at high risk for having or transmitting a specific genetic disorder.

Genetic screening involves testing an entire population to determine who among that group is at increased risk for a genetic disorder or for having a child with a genetic disorder.

Genome

The DNA contained in an organism or a cell, which includes both the chromosomes within the nucleus and the DNA in mitochondria.

Genome is a relatively new term coined to describe the collection of genes that are passed down from cell to cell and generation to generation in a given organism. A genome is the complete collection of genetic information, including the genes and the extra DNA used to package the DNA.

Related terms: cell, chromosome, DNA, mitochondrial DNA, nucleus

Genotype

Genotype is the genetic identity of an individual that does not show as outward characteristics.

A genotype, as described by a geneticist, is the inheritance pattern of a group of genetic markers for a particular individual. In short, it is a system of scoring part of the genome for the inheritance of variable genetic markers that are capable of describing any given individual's genetic makeup.

Related terms: genetic marker, phenotype

Germ Line

Inherited material that comes from the eggs or sperm and is passed on to offspring.

The cells of an organism —such as human or mouse— in which the sperm and oocytes divide are known as the germ line. During the early stages of development, these cells are separated from the remaining, or somatic, cells in the body in such a way that the germ line is treated differently. The germ line, then, generates the sperm and oocytes used during fertilization when producing the next generation.

Related terms: cell, inherited, somatic cells

Haploid

The number of chromosomes in a sperm or egg cell, half the diploid number.

Haploid refers to the number of copies of the genome in a particular cell. Normally, all of the cells in our body contain two copies of the genome—one inherited from each parent. But the cells of the reproductive system (that is, the sperm cells and the egg cells) carry only one copy. This copy is referred to as a haploid copy of the genome, and it contains the basic information that will eventually merge with another haploid cell to form the basics of the newborn.

Related terms: chromosome, diploid, genome

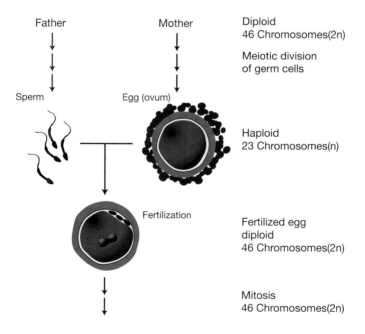

Father   Mother   Diploid
                                46 Chromosomes(2n)

                                Meiotic division
                                of germ cells

Sperm   Egg (ovum)

                                Haploid
                                23 Chromosomes(n)

           Fertilization

                                Fertilized egg
                                diploid
                                46 Chromosomes(2n)

                                Mitosis
                                46 Chromosomes(2n)

Haploinsufficiency

A situation in which the protein produced by a single copy of an otherwise normal gene is not sufficient to assure normal function.

> *Haploinsufficiency defines the result when the total level of gene product produced by a given gene equals only half the normal level. Three events can lead to haploinsufficiency: 1) one of the two copies of the gene is missing; 2) a mutation in the gene stops production of the message from that gene, or 3) the protein, or message, that is produced is unstable and is corrupted by the cell.*

Related terms: gene, protein

Hematopoietic Stem Cells

An unspecialized precursor cell that develops into a mature blood cell.

> *Hematopoietic stem cells are the most primitive of all blood cells in the circulation and have a long, indefinite life span. The word "hematopoietic" means related to the formation of blood cells. These cells are located in the bone marrow and are only present in a ratio of 1 in every 10,000 bone marrow cells. Hematopoietic stem cells are like mother cells in that they have the capacity to make new cells of different types. These cells are able to divide and self-renew, and to differentiate into mature cells of the blood such as erythrocytes, granulocytes, lymphocytes, and platelets. A problem in these cells is often the reason for bone marrow transplantation in an effort to repopulate a patient with stem cells that can make new blood cells.*

Related terms: cell

Hemophilia

A sex-linked inherited bleeding disorder that generally affects only males. The disorder is characterized by a tendency to bleed spontaneously or at the slightest injury. The disease is characterized by the absence of clotting factors.

A mutated gene on the long arm of the X chromosome causes the disease. Hemophilia is one of the more common sex-linked genetic diseases and has a long clinical history. One of the best-studied cases was that of the last Russian czar, Nicholas Romanov II, whose son was affected by hemophilia.

Related terms: inherited, protein

Heterozygous

Possessing two different forms of a particular gene, one inherited from each parent.

Every person has 46 chromosomes consisting of 23 pairs. In any one pair of chromosomes, for example in chromosome number 4, one member of that pair is inherited from the father and the other inherited from the mother. Genes can have variants in the population causing the gene to be slightly different in one individual than in another. If a person inherits two different variants of a gene on a pair of chromosomes (one from the father, one from the mother), that person is termed heterozygous for that gene, or is called a heterozygote.

Related terms: chromosome, gene, inherited

Inheritance of color vision

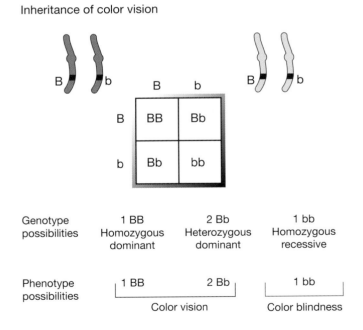

	B	b
B	BB	Bb
b	Bb	bb

Genotype possibilities	1 BB Homozygous dominant	2 Bb Heterozygous dominant	1 bb Homozygous recessive
Phenotype possibilities	1 BB 2 Bb		1 bb
	Color vision		Color blindness

Highly Conserved

A DNA sequence similar in several different kinds of organisms. Scientists regard these cross-species similarities as evidence that a specific gene performs some basic function essential to many forms of life and that evolution has therefore conserved its structure by permitting few mutations to accumulate in it.

When a gene is highly conserved, the sequence of this gene is quite similar in another species. For example, genes that are important in the development of the fly or the mouse brain are often very similar to genes that are important in the development of the human brain. By studying genes that are highly conserved in other species, such as the mouse or the fly, researchers often learn more about these genes in the human.

Related terms: gene, mutation

HIV/ AIDS

AIDS was first reported in 1981 in the United States and has since become a major epidemic, killing nearly 12 million people and infecting more than 30 million others worldwide. The disease is caused by the HIV virus, which destroys the body's ability to fight infections and certain cancers.

> *AIDS stands for acquired-immunodeficiency syndrome. It is an infectious disease, which is caused by HIV, which stands for human immunodeficiency virus. The virus attacks the cells in the body that normally help prevent infections, therefore the body loses the ability to fight off infections and even certain forms of cancer. AIDS is transmitted from person to person in body fluids that are cell-rich, such as blood or semen. This transfer may occur through sexual relations, contact with AIDS-contaminated blood, or the sharing of contaminated needles.*

Related terms: cell, retrovirus

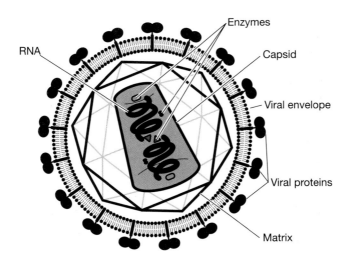

Homologous Recombination

Exchanging pieces of DNA during the formation of eggs and sperm. Recombination allows the chromosomes to shuffle their genetic material, increasing the potential of genetic diversity. Homologous recombination is also known as crossing over.

Homologous recombination is the substitution of a segment of DNA by another that is identical (homologous) or nearly identical. Homologous recombination occurs naturally during meiotic recombination and allows an exchange of alleles between chromosomes that may give rise to new character traits and combinations in the process. Homologous recombination can also be used as a laboratory technique to modify the sequence of a gene using gene targeting, and is sometimes referred to as "crossing over."

Related terms: chromosome, DNA

During Prophase 1 of Meiosis

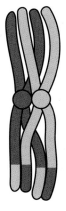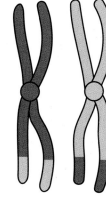

Crossing over
event

Result

Homozygous

Possessing two identical forms of a particular gene, one inherited from each parent.

Human beings ordinarily have 46 chromosomes grouped in 23 pairs. We actually have two copies of every gene in our genome. One copy of the gene is on one member of a pair of chromosomes, and the other copy of the gene is on the other member of the pair of a particular chromosome. Genes in humans can show variation or differences from one copy of the gene to another and are not always exactly alike. When an individual has two identical copies of a gene on his pair of a particular chromosome, he is "homozygous" for that gene or he is a homozygote.

Related terms: chromosome, gene, heterozygous, inherited

Inheritance of color vision

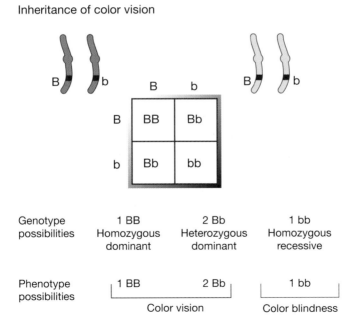

| Genotype possibilities | 1 BB Homozygous dominant | 2 Bb Heterozygous dominant | 1 bb Homozygous recessive |

| Phenotype possibilities | 1 BB   2 Bb | 1 bb |
| | Color vision | Color blindness |

Human Genome Project

An international research project to map each human gene and to completely sequence human DNA.

The Human Genome Project (HGP) has been described as the most ambitious scientific undertaking organized by humankind—even more so than splitting the atom or going to the moon. The HGP is an effort as a species to read life's instruction book. It is likely that in the future all of biology will be divided by what researchers did before they knew the genome and what researchers do after the genome's sequencing. The project is international in scope and will produce a map of the entire DNA sequence of human beings; all 3 billion base pairs. The HGP will also provide the DNA sequences of other organisms that are useful in terms of interpreting the meaning of the data. The HGP will provide a better understanding of human variation and susceptibility to disease, new technologies that will allow us to analyze situations we couldn't have analyzed otherwise, and new ways to manage, prevent, and cure illness.

Related terms: base pair, DNA, DNA sequencing, gene, mapping

Huntington's Disease

A degenerative brain disorder that usually appears in mid-life. Its symptoms, which include involuntary movement of the face and limbs, mood swings, and forgetfulness, get worse as the disease progresses. It is generally fatal within 20 years.

Huntington's disease (HD) normally affects middle-aged individuals who are 30 years old or older. The disease is characterized by "chorea," which is a loss of control of the hands and extremities. Choreac motion is the term often used to describe the dance-like movements that HD patients exhibit, and HD was once called Huntington's chorea. In addition to the chorea, patients undergo dementia, which is a loss of control of mental functioning. In 1993, researchers identified the gene defect that causes HD near the tip of chromosome 4. It is one of a number of triplet repeat disorders, or polyglutamine diseases, that causes the brain cells in HD patients to die. The area of the brain that is primarily affected is the basal ganglia. The identification of the gene has allowed scientists to examine the function of this gene in its normal and mutated forms. A number of animal models have been established that mimic the effects of HD. These animal models, primarily mice, are being used to screen for various therapeutic agents or drugs that may delay or stop the progression of the disease in humans.

Related terms: autosomal dominant, gene

Hybridization

Base pairing of two single strands of DNA or RNA.

Hybridization refers to one of the most common techniques used by geneticists and molecular biologists to study DNA and RNA. Hybridization exploits the process whereby complementary strands of DNA zip up together to form a double helix. Using probes made of single-stranded DNA molecules, molecular biologists can cause these to pair up in an appropriate way with their complementary DNA strands. Hybridization simply means the correct zipping up of complementary strands of DNA to form the double helix.

Related terms: DNA, double helix, ribonucleic acid (RNA)

Immunotherapy

The concept of using the body's own immune system to treat disease. Immunotherapy may also refer to the therapy of diseases caused by the immune system, allergies for example.

Many researchers believe future scientific breakthroughs will make it possible to employ this type of approach in the treatment of many diseases. For example, immunotherapy for the treatment of cancer might involve developing a cancer vaccine, or even immunizing a patient against his or her own tumor. In this way, such a treatment would enlist the power of the immune system to search the entire body and kill off cancer cells.

Related terms: cancer

In Situ Hybridization

The base pairing of a sequence of DNA to metaphase chromosomes on a microscope slide.

In situ hybridization is a technique that researchers use to determine the presence of a particular DNA or RNA sequence in a cell. By hybridizing a probe for a specific genetic sequence, researchers are able to look in the microscope and see the place in the cell (whether it is on the chromosome or in the cytoplasm of the cell) where that probe binds. This approach is useful in the mapping of genes and in the study of the expression of genes. It is also a key process in DNA fingerprinting.

Related terms: chromosome, DNA, metaphase, ribonucleic acid (RNA)

Inherited

Transmitted through genes from parents to offspring.

Inherited describes traits or characteristics that are transmitted through genes from parents to their offspring. Importantly, not all traits that are inherited are genetic, and not all traits that are genetic are inherited. Traits are commonly inherited from parent to child, but all of those are not directly due to genes. Some traits that run in a family are due to the environment that a family shares. For example, obesity can be due to family dietary habits and not due to genes.

Related terms: cancer, gene

Insertion

A type of chromosomal abnormality in which a DNA sequence is inserted into a gene, disrupting the normal structure and function of that gene.

An insertion is a type of mutation in a particular gene, or a chromosomal region, where a section of DNA is inserted in or near a gene and causes a change in the function of the gene. This change may act to turn the gene off, activate it inappropriately, or alter its function in some other way.

Related terms: chromosome, DNA, gene, mutation

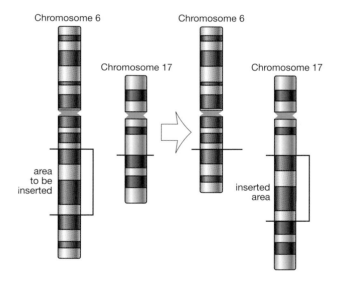

Intellectual Property Rights

Patents, copyrights, and trademarks.

> *Patents, copyrights, and trademarks encompass intellectual property rights. A patent, for example, is the right granted by the federal government to exclude others from making, selling, or using a claimed invention. Intellectual property rights in regard to genetic research involve issues such as the right to patent a gene or a gene sequence.*

Related terms: patent

Intron

A non-coding sequence of DNA that is initially copied into RNA, but is cut out of the final RNA transcript.

Genes are actually arranged on chromosomes as coding segments that end up specifying instructions for proteins. Between these sections of coding segments, or exons, are the introns, which do not code for portions of the proteins. Introns are spaces between the coding sections of genes on chromosomes that interrupt the coding sequence on a chromosome. Later in this process, these intron segments are cut out of the RNA before the message is used to make proteins. Introns are only found in higher organisms and are not present in one-celled organisms.

Related terms: DNA, gene, chromosome, genetic code, ribonucleic acid (RNA)

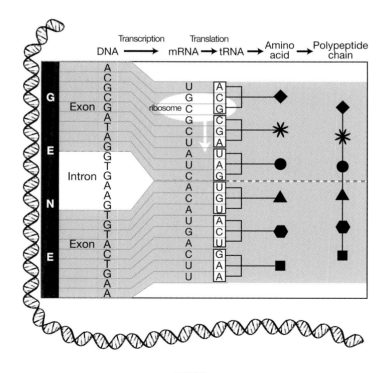

Karyotype

An individual's chromosomal complement, including the number of chromosomes and any abnormalities. The term also refers to a photograph of an individual's chromosomes.

A karyotype is the complete ensemble of all chromosomes present in a cell at the moment of cell division. Chromosomes are the compact structures of DNA within the nucleus of the cell, which contain the complete genetic information of the individual. The normal karyotype for a human is 23 pairs of chromosomes including one pair of sex chromosomes (the X and Y chromosomes). Karyotyping, which provides an actual picture of the chromosomes, can be a valuable tool in defining the diagnosis of some genetic diseases. For instance, Down's syndrome can be identified by karyotyping as it is evident as an abnormal number of chromosomes when blood cells are assessed in this manner.

Related terms: chromosome

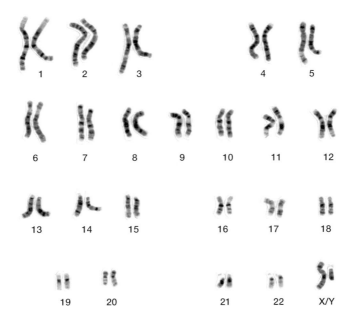

Knockout

Inactivation of specific genes. Knockouts are often created in laboratory organisms such as yeast or mice so scientists can study the knockout organism as a model for a particular disease.

> *Knockout usually refers to the inactivation of particular gene, as in "to knock out" a gene. It is more often used to refer to a specific type of animal model, usually a knockout mouse, that has had a gene, or genes "knocked out" so it can emulate a human disease. Knockouts are important for genetic researchers. By completely inactivating a particular gene function in a mouse, for example, researchers can understand the gene's role in the organism, and often in humans. This technique is commonly used in creating mouse models of human diseases such as cancer, obesity, and heart disease.*

Related terms: gene, locus

Leukemia

Cancer of the developing blood cells in the bone marrow. Leukemia leads to rampant overproduction of white blood cells (leukocytes); symptoms usually include anemia, fever, fatigue, weight loss, frequent infection, bleeding (including nose bleeds), and enlarged liver, spleen, and/or lymph nodes.

> *Leukemia is a cancer of the blood-forming cells which arises from cells in the bone marrow and in the lymph nodes. Leukemia is not a classic genetic disease, and does not pass from generation to generation. The disease is often associated with environmental effects such as chemical mutagens or exposure to radiation. Although leukemia is usually thought of as a childhood disease, it actually strikes ten times more adults each year than children.*

Related terms: cancer

Linkage

Genes and/or markers near one another on a chromosome; these tend to be inherited together.

Linkage is the association of genes and/or markers that lie near each other on a chromosome. "Linked" genes and markers tend to be inherited together. The simplest use of linkage occurs when a trait or a disease is inherited through a family and all of the people in the family also share a genetic marker. When the disease and the marker are inherited together in a family, a scientist can conclude that they must be closely located to each other on the chromosome. Linkage is the crucial first step in isolation of a gene. It allows a scientist to localize a gene to an area of DNA that may be 5-10 million base pairs out of the total 3 billion base pairs of DNA.

Related terms: base pair, chromosome, DNA, gene, genetic marker, inherited

Locus

Locus refers to the location of a specific gene on a chromosome; the plural is loci.

Each human chromosome contains thousands of genes. These genes are distributed on the 23 pairs of chromosomes. The locus is the particular place on a chromosome where a specific gene of interest is located. It's in a way, an address for the gene.

Related terms: chromosome, gene

LOD Score

A statistical estimate of whether two loci are likely to lie near each other on a chromosome and are therefore likely to be inherited together.

> When genetic information is passed from generation to generation, genes that are far from each other on a chromosome, or on separate chromosomes, are shuffled like a deck of cards. It is completely random as to what gene goes to child number one, versus child number two, and so on. If, however, two genes are very close together on a chromosome, they are usually inherited together, and they are not shuffled. A LOD score is a statistical test that determines whether two genes are located close to each other on the same chromosome. For example, a LOD score of 3 or more is generally taken to indicate that the two genes are quite close to each other on a chromosome.

Related terms: chromosome, gene, inherited, locus

Lymphocyte

A small white blood cell that plays a major role in defending the body against disease. There are two main types of lymphocytes: B cells, which make antibodies that attack bacteria and toxins, and T cells, which attack body cells themselves when they have been taken over by viruses or become cancerous.

> Lymphocytes are white cells found in the blood stream. They come in two varieties. T-lymphocytes fight viral infections and produce lymphocyte hormones that act on other cells to stimulate them. B-lymphocytes produce antibodies. Antibodies recognize bacteria and other invading pathogens so that they can be ingested by a third kind of cell in the body called a phagocyte. Phagocytes eat the invading organisms, as well as dead cellular debris, and act as the clean-up crew of the immune system.

Related terms: cancer, cell

Malformation

A structural defect often inherited in an organ or part of an organ that results from abnormal fetal development.

> Malformations come in many forms and vary widely in severity. Malformations can be normal. There are normal variations in human form and structure that are present in the healthy and normal population. One job of a geneticist is to separate malformations that cause disease and are part of a severe malformation syndrome from the normal malformation variants that are present in the healthy population.

Mapping

The process of deducing schematic representations of DNA. Three types of DNA maps can be constructed: physical maps, genetic maps, and cytogenetic maps, with the key distinguishing feature among these three types being the landmarks on which they are based.

> To geneticists, the process of making schematic representations of DNA is called mapping. Genetic mapping is very similar to the construction of geographical maps. Several different types of DNA maps can be constructed by geneticists. Genetic mapping is a central activity of the Human Genome Project.

Related terms: cytogenetic map, genetic map, Human Genome Project, physical map

Marker

A segment of DNA with a physical location on a chromosome whose inheritance can be followed, because DNA segments that lie near each other on a chromosome tend to be inherited together. Markers are often used as indirect ways of tracking the inheritance pattern of genes that have not yet been identified, but whose approximate locations are known.

> *A marker, or a genetic marker, is a segment of DNA with an identifiable physical location on a chromosome, and whose inheritance can be followed through a family. A marker can be a gene, or it can be a piece of DNA with no known function. Markers are important to geneticists because DNA segments that lie near each other on a chromosome tend to be inherited together. Markers are also the features of DNA that are used for genetic mapping to link genes to chromosomes, or to segments of chromosomes, as a first step in finding the position, identity, and function of a gene.*

Related terms: chromosome, DNA, gene, gene mapping, inherited

Melanoma

Melanoma is cancer of the cells in the skin that produce melanin, a brown pigment.

Melanoma develops in the pigment cells of skin, the eye, and moles, which are common in all adults. Melanoma, like all cancers, is a complex genetic disease. The rate of occurrence of melanoma is increasing more rapidly than that of any other cancer, with the exception of lung cancer in women. Currently, as many as 75 individuals per 100,000 will have this cancer. Although melanoma occurs equally in men and women, for still unknown reasons it has a better prognosis in women. Interestingly, an increased risk of death due to melanoma occurs as one approaches the equator; this has led to the suggestion that exposure to the sun is a major determinant of risk. Although light-skinned races are clearly at higher risk than pigmented races, the disease occurs across all racial backgrounds, where various forms of melanoma have been shown to be associated with a family predisposition.

Related terms: cancer, cell

Mendelian Inheritance

The manner in which genes and traits are passed from parents to children. There are three modes of Mendelian inheritance, autosomal dominant, autosomal recessive, and X-linked inheritance.

Humans have 23 pairs of chromosomes. An individual inherits one copy of each pair from his father and the other from his mother. In terms of Mendelian inheritance, you can inherit something either from your father or your mother. In the case of an inherited disease that is autosomal dominant, one parent with the disease will pass it on to the child because only one bad copy of the gene is necessary for this to occur. In contrast, with a disease that is autosomal recessive, both parents must pass on bad copies of the same gene to the child. In X-linked inheritance, the bad gene is on the X chromosome. Females have two X chromosomes and males have an X and a Y, which explains why X-linked diseases (such as muscular dystrophy) are seen only in males. If a boy receives a bad copy of the X chromosome from his mother, there is no other X chromosome to compensate. Whereas if a girl receives a bad copy of the X chromosome from her father and a good copy from her mother, the good copy compensates for the bad copy.

Related terms: autosomal dominant, chromosome, phenotype, recessive

Messenger RNA (mRNA)

mRNA is a template for protein synthesis. Each set of three bases, called codons, specifies a certain protein in the sequence of amino acids that comprise the protein. The sequence of a strand of mRNA is based on the sequence of a complementary strand of DNA.

Messenger RNA, or mRNA, lives up to its name. mRNA is a particular kind of RNA molecule which serves as a messenger that carries the particular gene sequence from the nucleus out to the cytoplasm where it is then translated into protein.

Related terms: gene, nucleus, protein, ribonucleic acid (RNA)

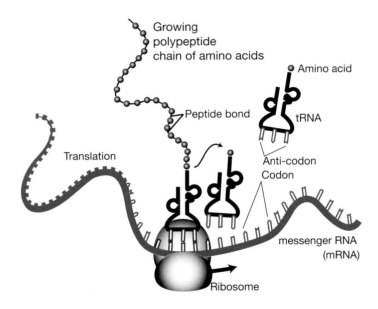

Metaphase

The phase during mitosis, or cell division, when the chromosomes align along the center of the cell. Because metaphase chromosomes are highly condensed, scientists use these chromosomes for gene mapping and identifying chromosomal aberrations.

Metaphase is a step in a process known as mitosis, which occurs when a cell divides. During mitosis, the DNA within the nucleus is divided into chromosomes, and compresses into thread-like structures visible under a microscope. The process of cell division has several steps at metaphase. The chromosomes line up together in a flat, plane-like manner, at which point they're ideally condensed and easy to see using a microscope. Metaphase is particularly important to cytogeneticists, who study chromosomes, because this is the stage at which they harvest the cells and prepare them for viewing. Cytogeneticists use the term metaphase chromosomes to indicate those chromosomes to be checked for abnormalities.

Related terms: cell, chromosome, gene, mapping

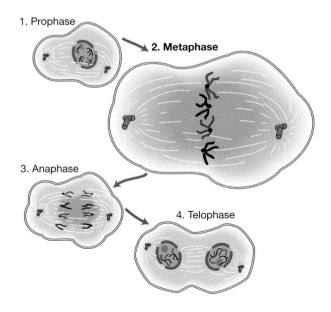

1. Prophase
2. Metaphase
3. Anaphase
4. Telophase

Microarray Technology

Microarray technology is a relatively new laboratory technique that researchers can use to see at how cells use their genetic information in their day-to-day functions. Through detailed, snapshot-like imagery, this technique allows researchers to view the inner workings of a cell, identify which genes in cells are actively making products, and determine whether cells are healthy. Patterns of unhealthy activity, which emerge from studying many of these snapshots, allow researchers to examine ways of treating unhealthy cells.

A new way of studying how large numbers of genes interact with each other and how a cell's regulatory networks control vast batteries of genes simultaneously. The method uses a robot to precisely apply tiny droplets containing functional DNA to glass slides. Researchers then attach fluorescent labels to DNA from the cell they are studying. The labeled probes are allowed to bind to complementary DNA strands on the slides. The slides are put into a scanning microscope that can measure the brightness of each fluorescent dot; brightness reveals how much of a specific DNA fragment is present, an indicator of how active it is.

Related terms: cell, DNA, gene

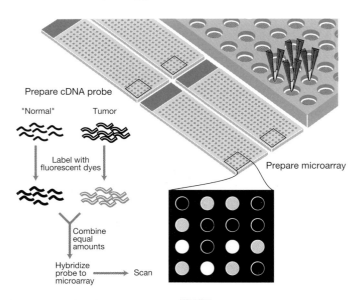

Mitochondrial DNA

The genetic material of the mitochondria, the organelles that generate energy for the cell.

Mitochondria are small structures (organelles) inside eukaryotic cells. Eukaryotic cells contain a specific nucleus, which holds most of the cell's DNA. Some human cells, such as the cells in the kidney or the brain, may contain hundreds of mitochondria. Known as biological workhorses, mitochondria supply the energy necessary for a cell to perform its duties. However, mitochondria also contain a different set of DNA (a different genome) than that found in the nucleus of the cell. Many scientists believe mitochondrial DNA to be the remnant DNA of a bacterium that invaded the cell very early in evolution and, over time, began to employ this bacterium's ability to produce large amounts of energy, while negating its detrimental effects. The bacterium's DNA became the DNA we now know as mitochondrial DNA. Mitochondrial DNA is a completely independent genome, has heritability, causes a genotype, and may cause disease when disrupted or mutated. Diseases associated with mitochondrial DNA mutations were first described in 1988 and result in numerous conditions including deafness, diabetes, and blindness.

Related terms: cell, DNA, genome, nucleus

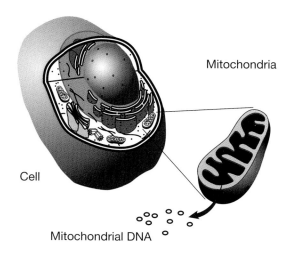

Mitochondria

Cell

Mitochondrial DNA

Mitochondrial DNA

Monosomy

Possessing only one copy of a particular chromosome instead of the normal two copies.

This loss of chromosomes leads to altered gene expression, usually an insufficiency of a gene product, which often leads to malformations, mental retardation, or other genetic diseases.

Related terms: cell, chromosome, gene expression, trisomy

Mouse Model

A laboratory mouse useful for medical research because it has specific characteristics that resemble a human disease. Strains of mice that have natural mutations similar to human mutations may serve as models of such conditions. Scientists can also create mouse models by transferring new genes into mice or by inactivating certain existing genes in them.

Mice play an important role in biomedical research because the mouse and human genomes are 80% to 85% percent similar. In addition, the majority of the biological functions that occur in mice are very similar to those that occur in humans, with only slight differences. This amazing similarity has enabled genetic researchers to model mice after conditions that occur in humans by creating mouse models of human disease. Researchers have learned to manipulate the mouse genome by inactivating (knocking out) certain genes, or by injecting (adding) DNA into the pronucleus of a mouse, a process known as transgenics. These techniques allow researchers to mimic a number of human ailments in a living mouse, enabling biologists to watch the onset and progression of diseases in a manner not possible in humans. Mouse models exist for hundreds of human diseases, including diabetes, cancer, and heart disease.

Related terms: animal model, gene, genome, mutation, transgenic

Multiple Endocrine Neoplasia

A rare inherited disorder that affects the endocrine glands, which release hormones into the bloodstream. The disorder, also known as Werner's syndrome, can cause multiple tumors in the parathyroid and pituitary glands and in the pancreas. These tumors are almost always benign, but they can cause the glands to become overactive and secrete abnormal levels of hormones. Those abnormal secretions, in turn, can cause a variety of medical problems, ranging from kidney stones and fatigue to fertility problems and life-threatening ulcers.

> *Multiple endocrine neoplasia type I (MEN I for short) is a rare, inherited disorder that affects the endocrine glands of the parathyroid, pituitary, and pancreas. A tumor in these glands produces an oversecretion of a hormone that results in certain medical problems such as kidney stones, fatigue, infertility, and at times life-threatening ulcers. Recent studies show that the MEN I gene is altered in families or individuals that developed these tumors at a later stage. The gene, discovered after an intensive 10-year search, is classified as a tumor suppressor gene.*

Related terms: gene, tumor suppressor gene

Mutation

A permanent structural alteration in DNA. In most cases, DNA changes either have no effect or cause harm, but occasionally a mutation can improve an organism's chance of surviving and passing the beneficial change on to its descendants.

A mutation describes a change in the DNA sequence that has a significant functional consequence, normally a negative consequence, particularly in human genetics. Geneticists most often reserve the term mutation for a change in the DNA sequence when they are confident cause and effect relate to an outcome of a disease or some other negative result. Mutation in other biological terms, where there are no human personal connotations, is a little more flexible, especially in a bacterium or yeast, where the consequences are not clearly harmful.

Related terms: birth defect, DNA

Niemann-Pick Disease, Type C

A disease that causes progressive deterioration of the nervous system by blocking movement of cholesterol within cells. The gene responsible for this disease, known as NPC1, is located on human chromosome 18.

Niemann-Pick disease, Type C is a childhood disorder that affects children aged 5 and older. The initial symptoms include an enlarged spleen and liver. The disease progresses to a point where loss of motor coordination leads to an unsteady gait, or ataxia, due to a loss of neurons in the cerebellum. Niemann-Pick Type C is a progressive disease that eventually leads to dementia, loss of speech capabilities, and the loss of muscle control and muscle tone. The defect is thought to occur due to the inability of the cell to metabolize cholesterol properly, leading to an excessive accumulation of cholesterol within the cells.

Related terms: cell, chromosome, gene

Nondirectiveness

The process by which genetic counselors advise clients toward a certain test or outcome, particularly related to child-bearing issues, without pressure or coercion.

> *Nondirectiveness is used by genetic counselors to avoid coercing or steering clients toward a certain test or a certain outcome, particularly with regard to pregnancy issues and decisions.*

Related terms: genetic counseling, risk communication

Nonsense Mutation

A single DNA base substitution resulting in a stop codon.

> *As part of the process of life in our cells, the genetic code is normally copied into amino acids that assemble into proteins. When a mutation in DNA causes one of those amino acids to be missing, the cell will stop making the protein by putting in a stop codon. This stop codon signals a halt to the protein-making process, resulting in a nonsense mutation. A nonsense mutation is a change in the genetic code where normally the code would tell the messenger RNA to make a protein, but due to a mutation, synthesis of the growing protein is stopped, and the protein is not made.*

Related terms: codon, DNA, genetic code, messenger RNA (mRNA), substitution

Northern Blot

A technique used to identify and locate mRNA sequences that are complementary to a piece of DNA called a probe.

A Northern blot, sometimes called a Northern analysis, is a technique that tests for the presence of mRNA, a molecule that serves as a messenger carrying a specific gene sequence from the nucleus to the cytoplasm. A Northern blot also allows researchers to obtain the size of a particular mRNA transcript.

Related terms: cDNA library, DNA, hybridization, messenger RNA (mRNA)

Nucleotide

One of the structural components, or building blocks, of DNA and RNA. A nucleotide consists of a base (one of four chemicals: adenine, thymine, guanine, and cytosine) plus a molecule of sugar and a phosphate group.

A nucleotide is the key structural component of DNA and RNA. The nucleotides in DNA contain four different bases—adenine, cytosine, guanine, and thymine. The nucleotides in RNA also contain four bases—adenine, cytosine, and guanine, but in the place of thymine there is a related base called uracil. Nucleotides are linked together to form a chain of DNA or RNA.

Related terms: base pair, DNA, genetic code, ribonucleic acid (RNA)

Deoxyribonucleic Acid (DNA)

Nucleus

The central structure within a cell that contains the chromosomes.

The DNA in human cells resides in a specific location called the nucleus. The nucleus is a spherical structure enclosed by a boundary membrane, and it contains the DNA and all the biochemical components that make the DNA function properly.

Related terms: chromosome, DNA

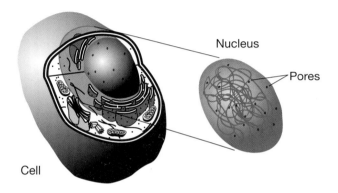

Nucleus

Pores

Cell

Oligonucleotide

A short sequence of single-stranded DNA or RNA. Oligonucleotides are often used as probes for detecting complementary DNA or RNA because they bind readily to complementary sequences.

An oligonucleotide (or an oligo) is a short strand of nucleotides, generally 10 or more strung together. The nucleotides are building blocks for DNA or RNA. When the string of nucleotides exceeds 50, it is often referred to as a fragment, whereas anything between 10 and 50 nucleotides is considered an oligo.

Related terms: DNA, ribonucleic acid (RNA)

Oncogene

A gene that is capable of causing the transformation of normal cells into cancer cells.

Cancer is a genetic disease often caused by the action, or inaction, of genes. Two types of genes cause cancer: oncogenes and tumor suppressor genes. An oncogene is a gene that makes a protein. These are normal proteins found in every cell, but occasionally the oncogene makes too much of this protein or makes a protein that is somewhat abnormal. When this occurs, the protein causes the cell to divide when it shouldn't, and abnormal cell division leads toward the formation of cancerous tumors.

Related terms: cancer, gene, protein, tumor suppressor gene

Oncovirus

A type of retrovirus that causes a cell to become cancerous.

> *An oncovirus is associated with causing a cell to become a cancer cell. When an oncovirus infects a cell, it introduces a gene that destroys normal growth regulation in the cell, which then grows out of control. Oncoviruses can be found in a variety of organisms from mice to humans.*

Related terms: cancer, cell, gene, retrovirus

p53

A gene that normally regulates the cell cycle and protects the cell from damage to its genome. Mutations in this gene cause cells to develop cancerous abnormalities.

> *p53 is a protein encoded by a particular gene, known as the p53 gene, which is located on chromosome 17 in humans. The p53 gene undergoes mutations, or changes, in human cancers. Of all of the genes that contribute to human cancer, p53 is the one that changes most frequently, causing this often-fatal disease. When unaltered, the p53 gene functions as a protector of the cell against damage to the genome. The p53 gene is also involved in monitoring the integrity of the genome. An appropriately functioning p53 gene activates when cells suffer damage. The gene serves either to stop the cell cycle (or replication of the cell) which allows DNA repair to occur, or, if there is a tremendous amount of damage, it sends a signal that causes the cell to self-destruct to avoid replication. When proper p53 function becomes impaired by mutations, the cells accumulate abnormalities and develop the genetic irregularities that lead to cancer.*

Related terms: cancer, cell, chromosome, gene, genome, inherited, mutation, tumor suppressor gene

Parkinson's Disease

A common progressive neurological disorder that results from degeneration of nerve cells in a region of the brain that controls movement. The first symptom of the disease is usually tremor of a limb, especially when the body is at rest.

Parkinson's disease is characterized by tremors and shakes in the extremities and it eventually progresses to difficulty with mobility and balance, and finally immobility. To date, researchers know that the disorder is associated with two abnormalities. The first is the loss of a specific group of neurons in the brain and the lack of a neuro-chemical, dopamine, in these neurons. One of the most effective treatments for Parkinson's disease has been the replacement of dopamine by administering the drug, L-dopa. The second is a mutation in the sequence of the human alpha synuclein gene, which appears to be responsible for the illness in a small percentage of families. This information is providing researchers new clues and new ideas in exploring the genetics of this debilitating illness.

Related terms: cell

Patent

When applied to genetics, a patent refers to the government regulations or requirements that confer the right or title to a gene to an individual or an organization. Genetics patents are granted only if there has been substantial human intervention in the discovery process.

In order to patent a gene, there are certain requirements set forth by the Patent and Trademark Office of the federal government. These are: (1) one cannot patent a product of nature unless there has been substantial intervention by man—cloning of genes proves substantial intervention; and (2) one must meet the three main criteria, which are novelty, utility, and non-obviousness. The gene must be new, therefore novel, it must have a defined use or function, and it must not be obvious based on information and the state-of-the-art research at the time the invention is claimed. If one can satisfy these criteria, there is a higher probability of obtaining a patent on the gene.

Related terms: gene, intellectual property rights

Pedigree

A simplified diagram of a family's genealogy that shows the relationships of family members to each other and how a particular trait or disease has been inherited.

A pedigree is a visual depiction of an extended family, constructed using connected circles and squares. The circles and squares are shaded to differentiate between those living and those deceased, as well as those with a particular disease and those without. This tool allows geneticists to visualize how family members are related: parents, grandparents, aunts and uncles, and siblings. A geneticist uses the pedigree to focus on health concerns or problems of each individual within that family in an attempt to identify visually whether a problem is familial (common over generations) or genetic.

Related terms: inherited

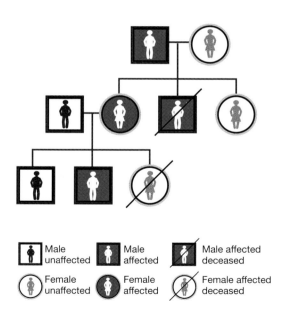

Male unaffected
Male affected
Male affected deceased
Female unaffected
Female affected
Female affected deceased

Peptide

Two or more amino acids joined by a peptide bond.

To define a peptide, we must understand that each cell is made of different components that are assembled into many different structures. One of the major components of these structures is protein. The proteins themselves are assembled from smaller units called amino acids. Amino acids are joined into chains; small chains of amino acids are defined as peptides. Longer chains of amino acids are known as polypeptides. Isolated or multiple polypeptides with specific functions are referred to as proteins.

Related terms: amino acids, protein

Structure of a tripeptide where R1,R2,and R3 represent 3 different amino acid side chains.

Phenotype

The observable traits or characteristics of an organism. Phenotypic traits are not necessarily genetic.

A phenotype describes a person's observable set of traits or characteristics; for example, hair color, weight, or the presence or absence of a disease. A phenotype is something that is always observable, whether clinically, in a laboratory, or through human and social interactions. A phenotype is not one's genetic makeup, but rather it is, in some sense, the expression, or the outcome, of one's genetic makeup as determined by one's genes and by the environment in which one grows and develops.

Related terms: gene, genotype

Physical Map

A chromosome map of a species that shows the specific physical locations of the genes and/or markers on each chromosome.

> *A physical map is a genetic map that depicts the positions of genetic landmarks along a stretch of DNA in a manner similar to a road map showing various points of interest along an interstate highway. In biology, these landmarks consist of genes, genetic markers, or anonymous sequences. Some physical maps are associated with an overlapping series of clones, such as YACs or BACs, that contain the corresponding DNA of that region. Thus, physical maps of chromosomes provide insight about the relative locations of genes and genetic markers across well-defined regions of the genome. Physical maps are often associated with clone sets that provide researchers organized access to the DNA from a specific chromosome. Physical maps are particularly important when searching for genes associated with a particular disease using the positional cloning method, and for sequencing all the chromosomes within a given genome.*

Related terms: bacterial artificial chromosome (BAC), chromosome, DNA, DNA sequencing, gene, genetic marker, Human Genome Project, mapping, positional cloning/functional cloning, yeast artificial chromosome (YAC)

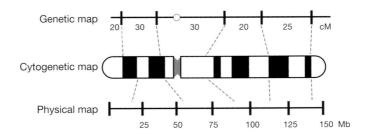

Polydactyly

An abnormality in which a person is born with more than the normal number of fingers or toes.

Polydactyly produces a malformation that affects the hands, or feet, in various ways. This may occur on the big toe or thumb-side of the limb, on the little toe, or pinky-side of the hand, or anywhere in between.

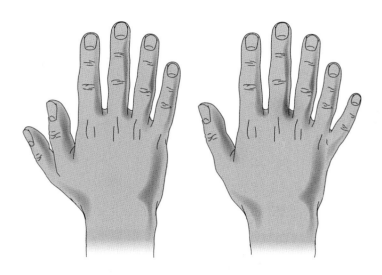

Polymerase Chain Reaction (PCR)

A fast, inexpensive technique for making an unlimited number of copies of any piece of DNA. Sometimes called "molecular photocopying," PCR has had an immense impact on biology and medicine, especially genetic research.

> *The powerful duplication ability of PCR allows genetic and molecular analysis using very small amounts of cells or tissues from any organism. PCR has been critically important for forensic analysis and in DNA fingerprinting. This is particularly valuable when people leave minute pieces of themselves, such as finger cells or hair cells, at a crime scene. With PCR, scientists can amplify or copy the DNA from those minute materials left behind and determine the source of those materials.*

Related terms: cell, DNA

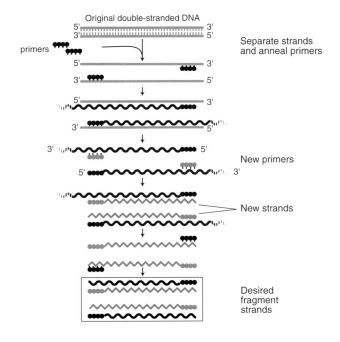

Polymorphism

A common variation in the sequence of DNA among individuals.

Polymorphism simply means more than one form. A polymorphism is a gene that exists in more than one version (or allele), and where the rare form of the allele can be found in more than 2% of the population. This is a somewhat arbitrary definition, and there is no biologic rationale for selecting the 2% cutoff. Alleles that are present in less than 2% of the population may still be polymorphisms. Researchers often consider the biology and medicine behind the gene to decide whether or not it is a polymorphism.

Related terms: DNA, genetic code

Positional Cloning / Functional Cloning

Two methods researchers employ to locate a gene responsible for a disease when little or no information about the biochemical basis of that disease is known.

Functional cloning can be used to find a gene when a researcher knows enough about a disease to estimate ehich genes are involved. This approach is based on information known about what "functionally" has gone awry in the disease. Unfortunately, not enough is known to utilize this strategy successfully for most diseases. For example, a lack of data prohibits the use of this method when studying diabetes or cancer. To find a gene by positional cloning, researchers collect families where the disease has occurred in numerous ancestors. DNA samples from these families are tested using genetic markers that encompass all human chromosomes; testing continues until researchers find a marker that tends to predict who is affected with the disease. Once this area of the genome is located, researchers search for a gene that has a different spelling in those affected from those unaffected. Positional cloning is complicated in that researchers begin with the entire genome (3 billion base pairs) when they're only looking for a mutation as subtle as a one-letter change, such as a T instead of a G.

Related terms: chromosome, cloning, gene, genetic marker

Primary Immunodeficiency

Primary immunodeficiency describes a group of very rare inherited diseases of the immune system. These diseases are the subject of intense genetic research aimed at better understanding the role of genes in the immune system, and immune system disorders in general.

> *One of these diseases, ADA (adenosine deaminase deficiency), was the first disease to successfully receive gene therapy as a treatment. ADA patients have defective cells responsible for the development of the immune system (T-cells which fight virus infections in the body and B-cells which fight bacterial infections). Historically, ADA resulted in death by overwhelming infection in the patient.*

Related terms: adenosine/deaminase (ADA) deficiency, severe combined immunodeficiency (SCID)

Primer

A short oligonucleotide sequence used in polymerase chain reaction (PCR).

Researchers frequently use the word primer in describing their research. A primer is often used when discussing PCR, which is a common laboratory technique that enables the researcher to make unlimited copies of any piece of DNA, and permits the analysis of very small amounts of cells or tissue. A primer is a short, single-stranded nucleic acid molecule used to prime a PCR. The primer finds its complement on the template strand once the PCR process is started.

Related terms: polymerase chain reaction (PCR)

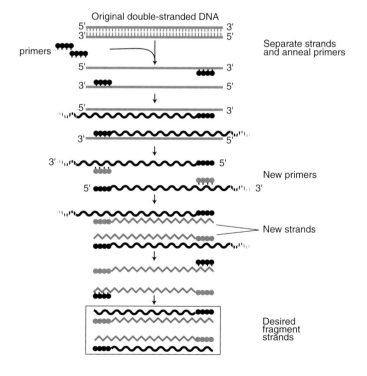

Promoter

The part of a gene that contains the information to turn the gene on or off. The process of transcription is initiated at the promoter.

A promoter doesn't make any proteins and acts much like the master switch location for the gene and is needed for the DNA to be turned into the RNA that makes the protein. All genes are broken up into three general parts. The first part, which is in front of the region that makes the protein, is the promoter. The promoter is where the gene's messenger RNA starts. Next comes the part of the gene that carries the information which is actually turned into the protein. The final part of the gene carries the information telling the messenger RNA where to stop. At a minimum, these three regions are needed in order to have a functioning gene: the promoter, the gene, and the messenger RNA stop.

Related terms: gene, DNA, protein

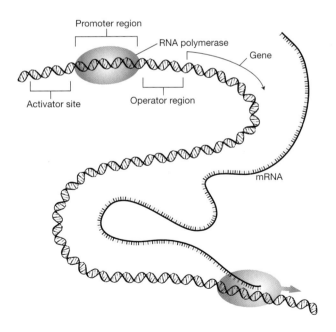

Pronucleus

The nucleus of a sperm or an egg cell before fertilization occurs.

A pronucleus is part of the nucleus in a fertilized egg just prior to the time in which the pronucleus from the sperm and the pronucleus from the egg fuse together to make a single nucleus. When researchers speak of a pronucleus, they are often speaking of that point in development in which both of these pronuclei are separated and distinct. This is the point at which the researcher can differentiate the two different pronuclei, and is valuable in the study of development, as well as in the making of animal models. For example, in making some mouse models of human disease, the DNA is inserted into the pronucleus of the fertilized egg of the mouse.

Related terms: chromosome, DNA, nucleus, transgenic

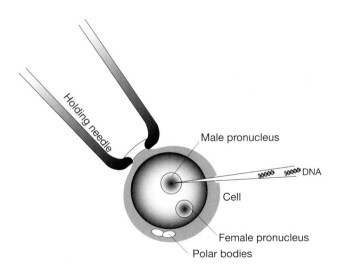

Prostate Cancer

Cancer of the prostate, a gland found only in men. The prostate surrounds the neck of the bladder and the urethra.

Cancer of the prostate gland, a small chestnut-shaped area located near the beginning of the urinary canal in men, accounts for one third of all cancers affecting men in the U.S. The incidence of prostate cancer shows a strong age, race, and geographic dependence. It is a disease of older men, and the incidence rate for men over 65 is 20 times greater than that of men 50–54. Less than 1% of all cases are diagnosed before the age of 40. Prostate cancer is very uncommon in certain ethnic populations such as Asians, whereas African Americans have extraordinarily high rates of prostate cancer. African Americans have the highest incidence and mortality rates known in the world. Most prostate cancers grow very slowly, and the disease can be treated, and even cured, if detected early. It is estimated that 5–10% of prostate cancers arise through a hereditary predisposition.

Related terms: cancer

Protease

A protein that digests other proteins.

> *Protease is a protein that cuts up other proteins in the human body. Proteases are enzymes important in many forms of regulation in the cell. There are many types of proteases, all having specific types of proteins that they work best at breaking down so the protein can then be assembled and reused in the cell.*

Related terms: cell, enzyme, protein

Protein

A large complex molecule made up of one or more chains of amino acids. Proteins perform a wide variety of activities in the cell.

Proteins are large molecules that have many important functions in the human body. For instance, the enzymes that help you digest your food are proteins. Similarly, the muscles that help you move your arms are made of specific muscle proteins. The hair on your head is made up of a different protein. Proteins are also very important to geneticists. Genes code for proteins, meaning that proteins are made from the sequence of genes. When these proteins have an altered sequence, they can cause disease. Likewise, when certain proteins are not present, or when there is too much of a protein (such as in gene amplification), proteins can also cause disease.

Related terms: amino acids, cell, gene, gene amplification

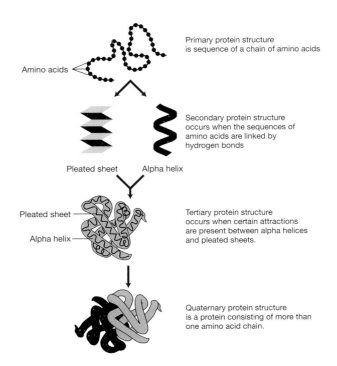

Amino acids

Primary protein structure is sequence of a chain of amino acids

Pleated sheet   Alpha helix

Secondary protein structure occurs when the sequences of amino acids are linked by hydrogen bonds

Pleated sheet

Alpha helix

Tertiary protein structure occurs when certain attractions are present between alpha helices and pleated sheets.

Quaternary protein structure is a protein consisting of more than one amino acid chain.

Pseudogene

A sequence of DNA that is very similar to a normal gene but that has been altered slightly so it is not expressed. Such genes were probably once functional but over time acquired one or more mutations that rendered them incapable of producing a protein product.

> *A pseudogene is a defective copy of a gene. Each human chromosome has much information that does not productively code for a protein. Although this information may appear as a functioning gene, upon closer examination of the genetic code, it turns out to not be a working gene at all. While 90% of our genes may be able to code for a protein, the remainder are nonfunctional. These remaining genes are called pseudogenes.*

Related terms: DNA, gene, protein

Recessive

A genetic disorder that appears only in patients who have received two copies of a mutant gene, one from each parent.

Recessive refers to the way in which a trait, or mutation, is passed from the parents to the child. A recessive mutation requires a double dose, or two copies, of the mutated gene be passed to the child (one copy from the mother and one copy from the father) in order for the defect to appear in the child. If the child receives only one mutated gene and one normal gene, the recessive trait will not manifest itself in the offspring.

Related terms: cancer, gene, heterozygous, mutation

Inheritance of color vision

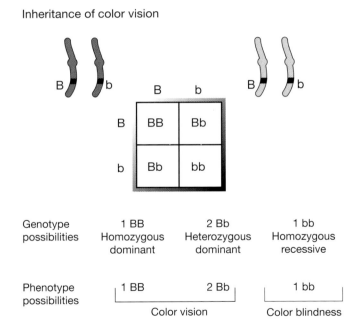

| Genotype possibilities | 1 BB Homozygous dominant | 2 Bb Heterozygous dominant | 1 bb Homozygous recessive |

| Phenotype possibilities | 1 BB | 2 Bb | | 1 bb |
| | | Color vision | | Color blindness |

Recombinant DNA

A variety of techniques that molecular biologists use to manipulate DNA molecules in order to study the expression of a gene.

Recombinant DNA is a catchall phrase to describe a variety of techniques that molecular biologists use to manipulate DNA molecules. In this process researchers take a DNA molecule from one organism, perhaps a virus, a plant or bacteria, bring it into the laboratory and manipulate it, and then place it in another organism. This technique is often done to study the expression of a certain gene and, in some cases, in an attempt to treat human genetic disease.

Related terms: bacteria, chromosome, DNA, gene, gene expression

Restriction Enzymes

Enzymes that recognize a specific sequence of double-stranded DNA and cut the DNA at that site. Restriction enzymes are often referred to as molecular scissors.

Restriction enzymes are proteins that have a property that allows them to stick to DNA, and they do so at specific locations. Restriction enzymes actually cut the DNA molecule at a specific location or at a specific sequence of DNA bases. Restriction enzymes are used widely in molecular biology and form the backbone of recombinant DNA technology—that is, the ability to cut apart different pieces of DNA and stick them back together again.

Related terms: DNA, enzyme

Restriction Fragment Length Polymorphism (RFLP)

RFLPs are genetic variations in the length of DNA sequence fragments produced when researchers manipulate, or cut, DNA using a restriction enzyme. Restriction enzymes cut DNA at very specific sequences of base pairs.

> RFLPs were once commonly used by genetic researchers to look for genetic variation in the human DNA sequence, or polymorphisms. They are also useful as markers in genetic maps. In its heyday, around the 1980s, the RFLP was the researcher's best tool to look at genetic variation in DNA. However, with the advent of PCR, the way researchers look for DNA variation changed significantly. Today most researchers no longer use RFLPs to study human variation. Instead, they look at the actual genetic sequence in a more precise way using research approaches such as SNPS and microsatellites.

Related terms: DNA, enzyme, linkage, polymorphism, single nucleotide polymorphisms (SNPs)

Retrovirus

A type of virus that contains RNA as its genetic material. The RNA of the virus is translated into DNA, which inserts itself into an infected cell's own DNA. Retroviruses can cause many diseases, including some cancers and AIDS.

> A retrovirus is a potent disease-causing virus. As part of their attack on the human body, retroviruses access the chromosomes of the cells they invade. Scientists have learned how to make these viruses harmless, and to use their ability to penetrate into the human cells to actually transfer beneficial genes for use in human gene therapy.

Related terms: cancer, cell, DNA, HIV/AIDS, ribonucleic acid (RNA)

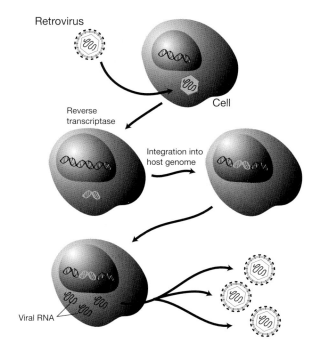

Ribonucleic Acid (RNA)

A chemical similar to a single strand of DNA. In RNA, the letter U (uracil) is substituted for T (thymine) in the genetic code. RNA delivers DNA's genetic message to the cytoplasm of a cell where proteins are made.

RNA is ribonucleic acid, a chemical very similar to a single strand of DNA (DNA is usually present in double strands). RNA delivers DNA's genetic message to the cytoplasm of a cell where proteins are made. RNA that carries this genetic message is called messenger RNA (or mRNA), and is one of the most important forms of RNA in a cell. Other forms of RNA actually are structural and form the machinery that converts RNA into proteins.

Related terms: cell, DNA, genetic code, messenger RNA (mRNA), protein

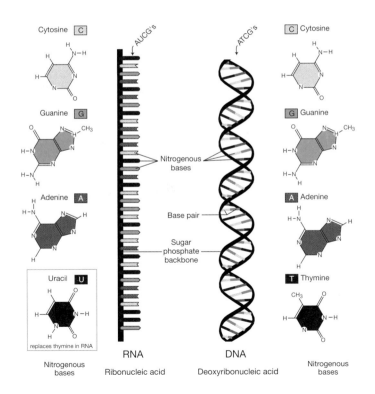

Ribosome

Cellular organelle that is the site of protein synthesis.

Ribosomes are particles that occur in the cytoplasm of cells. They are composed of a mixture of RNA and protein, and can be characterized as little factories that contain the equipment necessary to read out the message in messenger RNA and translate it into protein.

Related terms: messenger RNA (mRNA), protein, ribonucleic acid (RNA)

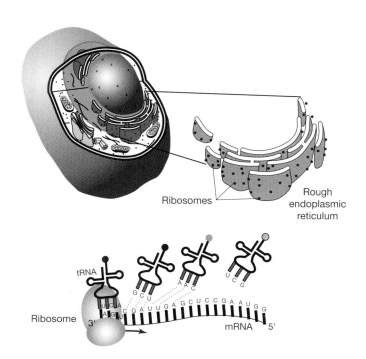

Ribosomes

Rough endoplasmic reticulum

tRNA

Ribosome

mRNA

Risk Communication

An educational process through which a genetic counselor attempts to interpret how a genetic condition is inherited and the chance that it might be passed on to children.

> Genetic counselors are key clinical professionals whose roles include enabling individuals and families to understand genetic conditions, genetic tests, diagnosis of a genetic disease, and similar life circumstances involving medical genetics. Much of this process is aimed at helping individuals understand their personal situation so they are able to make an informed choice regarding issues such as testing, treatment, parenting, and a host of other life situations. Through risk communication, the counselor typically seeks to help the individual in the areas specific to the inheritance of a genetic disorder. This may involve better understanding of family pedigrees and inheritance patterns (such as recessive and dominant), and a more specific understanding of the condition of interest.

Related terms: nondirectiveness, genetic counseling

Sequence-tagged Site (STS)

A short DNA segment that occurs only once in the human genome and whose exact location and order of bases are known. Because each is unique, STSs are helpful for chromosome placement of mapping and sequencing data from many different laboratories. STSs serve as landmarks on the physical map of the human genome.

> Physical maps reflect the positions of landmarks across a stretch of DNA. One type of landmark is the sequence-tagged site, or STS. An STS is a short, unique DNA sequence that can be specifically detected by PCR. Because PCR is very efficient and powerful, there are many advantages of having these DNA landmarks detectable by PCR. As a result, the initial physical maps of human chromosomes constructed in the Human Genome Project are largely being based on STSs.

Related terms: base pair, chromosome, DNA, genome, Human Genome Project, mapping, physical map, polymerase chain reaction (PCR)

Severe Combined Immunodeficiency (SCID)

A disease affecting the immune system. SCID is fatal if affected individuals do not receive bone marrow transplants.

> *SCID is a very rare inherited disease of the immune system. It is called combined immunodeficiency because both B and T lympho-cytes are compromised with a SCID mutation. This leaves the person with a SCID mutation at great risk for developing severe infections. People affected with SCID have very limited life spans unless they are treated with bone marrow transplantation.*

Related terms: antibody, bone marrow transplantation

Sex Chromosomes

One of the two chromosomes that specify the sex of an organism. Humans have two kinds of sex chromosomes, one called X and the other Y. Normal females possess two X chromosomes and normal males one X and one Y.

> *Sex chromosomes refer to the two chromosomes that identify whether a person is a male or a female. Humans have 23 pairs of chromosomes, making a total of 46. The first 22 pairs are called autosomes and are numbered 1 – 22 according to physical size (with 1 being the largest). The 23rd pair comprises the sex chromosomes. Females have two X chromosomes, and males have an X and a Y chromosome.*

Related terms: gene, mutation

Shotgun Sequencing

An approach used to decode an organism's genome by shredding it into smaller fragments of DNA that can be sequenced individually. The sequences of these fragments are then ordered, based on overlaps in the genetic code, and finally reassembled into the complete sequence. The "whole genome shotgun" method is applied to the entire genome all at once, whereas the "hierarchical shotgun" method is applied to large, overlapping DNA fragments of known location in the genome.

Shotgun sequencing means breaking a piece of DNA into smaller pieces and then assembling, or piecing it back together, by reading the A, C, G, and Ts. The "whole genome shotgun" approach involves mincing an entire genome all at once, and piecing it back together by matching overlaps in genetic code. This method was used successfully to sequence the fruit fly genome, and more recently, the human genome. Another approach to shotgun sequencing of the genome is more hierarchical and involves an intermediate step of breaking the genome down into more manageable pieces, such as 200,000 base pairs before mincing that particular section.

Related terms: DNA sequencing, genetic code, genome, Human Genome Project, mapping

Sickle-cell Anemia

A blood condition seen most commonly in people of African ancestry. The disorder is caused by a single base pair change in one of the genes that codes for hemoglobin, the blood protein that carries oxygen. This mutation causes the red blood cells to take on a sickle shape, rather than their characteristic donut shape. Individuals who suffer from sickle-cell disease are chronically anemic and experience significant damage to their heart, lungs, and kidneys.

Sickle-cell anemia is a serious blood condition seen most commonly in people of African ancestry, but also occurring in people of Arabic, Greek, Italian, Latin American and Native American descent. Sickle-cell disease is a recessive condition, which means that a person must inherit an altered gene from each parent and, when two altered genes are present, sickle-cell disease is the result. A single nucleotide change, or a mutation, in the beta-globin gene causes the disease, which is located on human chromosome 11. Beta-globin is a protein. It is part of the hemoglobin, which is the pigment in red blood cells, the cells that carry oxygen to all cells in our body. In a normal red blood cell containing normal beta-globin, the cell looks like a Frisbee and acts like a water balloon. Its contents can be squeezed easily back and forth, allowing the red blood cell to pass freely through small blood vessels. Sickle-cell disease creates an abnormal beta-globin protein, which may lead to sickle-shaped red blood cells. These sickle-shaped red blood cells look like a quarter moon, instead of a Frisbee, and act like a balloon filled with ice chips. Sickle cells do not squeeze very well through small blood vessels, and this causes damage to the vascular wall throughout the body. Damage to the vessels results in strokes in about a third of these patients before they reach the age of 10. Another third of these patients die of emphysema, which is caused by damage to the vessels of the lung, before age 20. Patients that make it past age 20 are anemic throughout life, and are often in pain as the sickle-shaped red blood cells block the vessels that supply blood to the fingers, toes, and lower back.

Related terms: base pair, gene, cell, protein, recessive

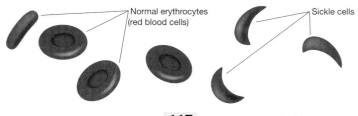

Normal erythrocytes (red blood cells)

Sickle cells

Single Nucleotide Polymorphisms

Common, but minute, variations that occur in human DNA at a frequency of one every 1,000 bases. These variations can be used to track inheritance in families. SNP is pronounced "snip".

> *Single nucleotide polymorphisms, or "snips", are the major genetic variations between individuals in the human genome. Most of the genetic variation, or genetic differences, between people are of this single-letter type where there might be a "G" in one person and a "T" in the other in the genetic code. Researchers believe there are about 10 million snips common in the human species, or 10 million positions along the 3 billion base pair human genome that are common variations. Snips are important because they allow researchers to track inheritance in families. Some snips are functionally important, with a small number playing a role in the susceptibility of diseases. Researchers are developing a large catalog of snips in order to help uncover the causes of major diseases such as diabetes, heart disease, and schizophrenia.*

Related terms: DNA, genetic code, inherited, nucleotide, polymorphism

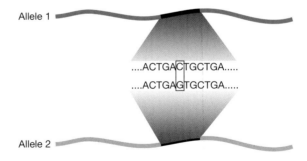

Somatic Cells

All body cells, except the reproductive cells.

> Somatic refers to body. The most frequent use of somatic in genetics is to distinguish between a somatic mutation and a germ-line mutation. A somatic mutation occurs in a cell of the body and will not be passed on to future generations. A somatic mutation, however, might yet cause some problems such as cancer in the individual in whom it occurs. A germ-line mutation occurs in a cell that will be passed on to the next generation.

Related terms: cell

Spectral Karyotyping

A visualization of all of an organism's chromosomes together, each labeled with a different color. This technique is useful for identifying chromosome abnormalities.

> Spectral karyotyping (or SKY) is a molecular, cytogenetic technique that allows researchers to view all 23 human chromosome pairs in vivid, specifically labeled, different fluorescent colors on a computer screen. The different colors of the chromosomes are based in the spectral analysis of the fluorescent signals. The resulting tool affords an improved way for clinicians and researchers to view human chromosomes over the older method of black-and-white karyotyping. It is expected this technique will greatly benefit the identification and diagnosis of chromosomal aberrations in prenatal cytogenetics, and the diagnosis and detection of chromosomal aberrations in cancer cells.

Related terms: chromosome, karyotype

Substitution

Replacement of one nucleotide in a DNA sequence by another nucleotide or replacement of one amino acid in a protein by another amino acid.

Geneticists use the word substitution to refer to specific types of changes in the genetic code. A substitution is the replacement of one nucleotide in a DNA sequence by another nucleotide. It can also apply to the replacement of an amino acid in a protein that is different from the normal amino acid that is present in that position. These are forms of mutations that either can be inherited within a family or can arise newly in an individual.

Related terms: amino acids, inherited, mutation, nucleotide, protein

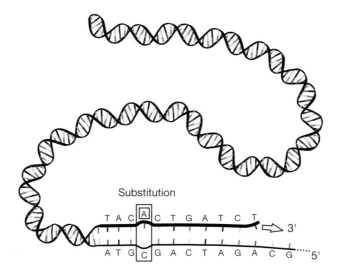

Substitution

Transgenic

An experimentally produced organism in which DNA has been artificially introduced and incorporated into the organism's germ line, usually by injecting the foreign DNA into the nucleus of a fertilized embryo.

Transgenic is a technical term referring to either a plant or an animal where a piece of DNA not normally found in that plant or animal is introduced into its cells. Transgenic usually refers to the introduction DNA into the germ line of cells of an animal model such as a mouse. Germ-line cells are the cells that pass on to the animal's offspring and are of importance in many genetic studies. Transgenic animals allow researchers to study the expression of introduced genes in the context of a whole animal. This allows researchers to observe what this extra gene does to a whole living organism, instead of looking at one cell in a tissue culture dish. This is important because many human diseases (cancer and heart disease, for example) do not just affect a single cell-type and, instead, affect the interactions between many different cell types. These processes cannot be studied in a cell culture environment and must be examined in the context of an whole organism—often the transgenic mouse.

Related terms: cell, DNA, gene, germ line, nucleus

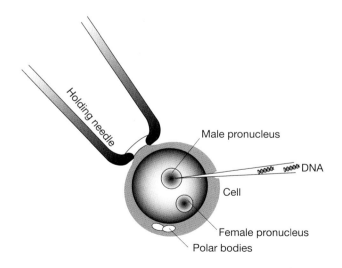

Translocation

Breakage and removal of a large segment of DNA from one chromosome, followed by the segment's attachment to a different chromosome.

Translocation occurs when a portion of one chromosome breaks away and affixes itself to another chromosome. Most often this happens in pairs; for example, a portion breaks away from both chromosome 1 and chromosome 3, and those pieces exchange position and affix themselves to the opposite chromosome. The portion from chromosome 1 attaches to chromosome 3 and vice versa. The result is a translocation between chromosome 1 and chromosome 3. Translocations can also occur on a single chromosome; this is known as an inversion. Frequently, in translocations or inversions, the break point is within the gene, damages the function of that gene, and can lead to disease.

Related terms: chromosome, DNA, gene

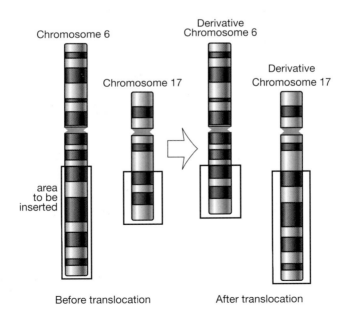

Before translocation   After translocation

Trisomy

Possessing three copies of a particular chromosome instead of the normal two copies.

> *Trisomy is a situation in which a person's cell, or all of a person's cells, have three copies of a particular chromosome instead of the usual two copies. This increase in copy number of a gene, or genes, leads to abnormalities of gene expression, causing malformation or other types of genetic diseases. The most common example of this is Trisomy 21, more commonly known as Down's syndrome.*

Related terms: cell, chromosome, gene, gene expression

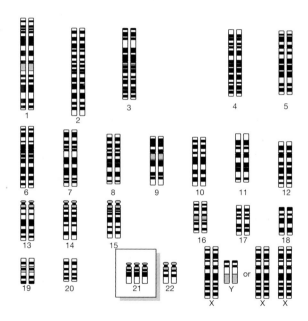

Tumor Suppressor Gene

A protective gene that normally limits the growth of tumors. When a tumor suppressor is mutated, it may fail to keep a cancer from growing. BRCA1 and p53 are well-known tumor suppressor genes.

A tumor suppressor gene belongs to the family of genes involved in the formation of tumors. Unlike oncogenes, which promote cell growth, the normal role of a tumor suppressor gene is to control cell growth. A malfunctioning tumor suppressor gene allows cells to divide abnormally and form a tumor. Examples of tumor suppressor genes include the gene involved in the eye tumor called retinoblastoma, and also the genes involved in the predisposition to breast cancer: the BRCA1 and BRCA2 genes. Gene function is often lost in these tumors, which aids tumor formation.

Related terms: BRCA1/BRCA2, cancer, gene, oncogene, p53

Vector

An agent, such as a virus or a small piece of DNA called a plasmid, that carries a modified or foreign gene. When used in gene therapy, a vector delivers the desired gene to a target cell.

A vector is a transport mechanism for delivering a gene into the DNA of an existing cell with the purpose of transferring genetic material. The term is commonly used in gene therapy applications when researchers are describing the system, or process, by which a gene moves from point A to point B. Common vectors important in gene therapy include a class of viruses called retroviruses.

Related terms: cell, DNA, gene, gene therapy, gene transfer

Yeast Artificial Chromosome (YAC)

Extremely large segments of DNA from another species incorporated into the DNA of yeast. YACs are used to clone up to one million bases of foreign DNA into a host cell, where the DNA is propagated along with the yeast cell's other chromosome.

> Large pieces of DNA can be engineered to contain the components necessary to allow them to be reproduced as artificial chromosomes in yeast. In this way, individual large DNA fragments that are a million bases long (or longer) can be cloned in yeast as yeast artificial chromosomes, or YACs. The cloned DNA is then bred as if it were just another chromosome in the yeast cell. These YACs are then used in the making of a physical map of a chromosome, and have specific advantages due to their size. In terms of a jigsaw puzzle, it is easier to assemble a small number of large pieces than a larger number of smaller pieces. The same is true for making physical maps of chromosomes. YACs provide big pieces of the chromosome requiring fewer to complete the puzzle of a chromosome physical map. As a result, YACs are particularly helpful in piecing together the human genome, where they have been used to construct complete physical maps of all the human chromosomes.

Related terms: base pair, chromosome, cloning, DNA, Human Genome Project

About the Authors

JEFFRE L. WITHERLY has been creating ways to help the public better understand important medical and scientific principles for over twenty years. His career highlights include hosting a syndicated radio health show, developing Internet-based learning tools such as the Talking Glossary of Genetics, and creating one of New England's first televised Allergy Indexes. He currently directs science education at the National Human Genome Research Institute.

GALEN P. PERRY's experience in communicating medicine and science to the public comes from having worked in the medical and research field for over a decade. He is currently Deputy Director of the Office of Science Education at the National Institutes of Health's National Human Genome Research Institute in Bethesda, Maryland, where he continues to create novel ways to explain complex scientific terms in a simple, straightforward manner.

DARRYL L. LEJA is a graduate of the University of Michigan's MFA program in medical illustration. His illustrations have appeared in numerous scientific journals and a number of the country's leading newspapers and news magazines. He currently directs the scientific illustration unit at the National Human Genome Research Institute, NIH, where he focuses on conceptual visualization for science education and research, including three-dimensional modeling.